BIG BOOK OF MATH

PRACTICE PROBLEMS

Multiplication & Division

by

Stacy Otillio & Frank Otillio

CLAY MAZE.com

EMPOWERING CHILDREN
FOR A SUCCESSFUL FUTURE

TABLE OF CONTENTS

Section 1
Multiplication Facts 1

Section 2
Find the Missing Multipliers (Multiplication Facts) 13

Section 3
Division Facts 25

Section 4
Find the Missing Divisors (Division Facts) 37

Section 5
Multiplication (2 Digits x 1 Digit) 49

Section 6
Multiplication (4 Digits x 1 Digit) 61

Section 7
Multiplication (3 Digits x 2 Digits) 73

Section 8
Division (3 Digits / 1 Digit) 85

Section 9
Division (5 Digits / 1 Digit) 97

Section 10
Division (4 Digits / 1 Digit) with Remainders 109

TABLE OF CONTENTS

Section 11
Division (4 Digits / 2 Digits) 121

Section 12
Multiplication by 10, 100, 1000 133

Section 13
Find the Missing Multipliers: 10, 100, 1000 145

Section 14
Multiplication by 10, 100, 1000 with Decimals 157

Section 15
Find the Missing Multipliers: 10, 100, 1000 with Decimals ...169

Section 16
Division by 10, 100, 1000 181

Section 17
Find the Missing Divisors: 10, 100, 1000 193

Section 18
Division by 10, 100, 1000 with Decimals 205

Section 19
Find the Missing Divisiors: 10, 100, 1000 with Decimals 217

Solutions
Solutions to Problems (Sections 1-19) 229

SECTION 1

MULTIPLICATION FACTS

11 worksheets
20 problems per sheet

Name ____________________ Date ____________

Multiplication Facts

Multiply.

1) 6 × 2	2) 2 × 9	3) 3 × 7	4) 8 × 2	5) 2 × 1
6) 8 × 9	7) 12 × 7	8) 3 × 3	9) 9 × 4	10) 8 × 5
11) 6 × 0	12) 7 × 5	13) 10 × 5	14) 9 × 2	15) 10 × 6
16) 11 × 9	17) 5 × 8	18) 10 × 11	19) 12 × 8	20) 10 × 9

Name ______________________ Date ______________

Multiplication Facts

Multiply.

1) 8 x 7	2) 2 x 9	3) 12 x 11	4) 11 x 10	5) 8 x 9
6) 7 x 9	7) 2 x 5	8) 6 x 7	9) 7 x 0	10) 6 x 4
11) 6 x 1	12) 10 x 3	13) 2 x 8	14) 5 x 3	15) 11 x 2
16) 3 x 3	17) 12 x 7	18) 7 x 7	19) 7 x 2	20) 9 x 9

Name ______________________ Date ______________

Multiplication Facts

Multiply.

1) 11 x 3	2) 10 x 12	3) 12 x 5	4) 8 x 5	5) 8 x 9
6) 12 x 6	7) 3 x 5	8) 4 x 1	9) 11 x 4	10) 4 x 4
11) 5 x 9	12) 5 x 5	13) 5 x 0	14) 8 x 7	15) 8 x 8
16) 10 x 3	17) 10 x 8	18) 2 x 9	19) 6 x 2	20) 4 x 5

Name ______________________________ Date ______________________

Multiplication Facts

Multiply.

1) 7 x 6	2) 5 x 1	3) 2 x 7	4) 7 x 5	5) 9 x 7
6) 4 x 8	7) 7 x 8	8) 2 x 6	9) 10 x 12	10) 11 x 3
11) 11 x 5	12) 6 x 4	13) 9 x 9	14) 8 x 0	15) 8 x 4
16) 6 x 7	17) 5 x 8	18) 3 x 3	19) 4 x 5	20) 5 x 2

Name ______________________________ Date ____________________

Multiplication Facts

Multiply.

1) 2 x 6	2) 3 x 7	3) 7 x 1	4) 10 x 11	5) 6 x 0
6) 6 x 4	7) 6 x 3	8) 10 x 2	9) 12 x 6	10) 11 x 4
11) 11 x 5	12) 5 x 6	13) 7 x 6	14) 4 x 2	15) 3 x 8
16) 9 x 3	17) 7 x 2	18) 12 x 7	19) 6 x 2	20) 4 x 4

Name ____________________ Date ____________________

Multiplication Facts

Multiply.

1) 5 x 8	2) 2 x 1	3) 4 x 5	4) 6 x 0	5) 7 x 6
6) 3 x 3	7) 10 x 2	8) 2 x 6	9) 8 x 4	10) 2 x 5
11) 7 x 4	12) 8 x 5	13) 5 x 3	14) 7 x 5	15) 10 x 9
16) 5 x 4	17) 9 x 5	18) 6 x 2	19) 6 x 7	20) 11 x 10

Name ____________________ Date ____________

Multiplication Facts

Multiply.

1) 8 x4	2) 11 x0	3) 8 x6	4) 8 x2	5) 9 x1
6) 3 x3	7) 6 x2	8) 2 x6	9) 2 x3	10) 12 x8
11) 5 x4	12) 11 x10	13) 12 x11	14) 11 x6	15) 7 x2
16) 4 x4	17) 12 x12	18) 9 x8	19) 5 x7	20) 11 x5

Name ______________________ Date ______________

Multiplication Facts

Multiply.

1) 2 x 7	2) 6 x 6	3) 10 x 11	4) 2 x 3	5) 3 x 1
6) 8 x 7	7) 4 x 3	8) 11 x 5	9) 10 x 9	10) 2 x 5
11) 12 x 3	12) 12 x 10	13) 7 x 3	14) 12 x 12	15) 4 x 6
16) 8 x 4	17) 9 x 5	18) 9 x 4	19) 12 x 9	20) 2 x 0

Name ______________________________ Date ____________________

Multiplication Facts

Multiply.

1) 9 x 2 = ____

2) 2 x 8 = ____

3) 11 x 10 = ____

4) 11 x 12 = ____

5) 10 x 11 = ____

6) 8 x 0 = ____

7) 3 x 3 = ____

8) 10 x 12 = ____

9) 5 x 7 = ____

10) 8 x 7 = ____

11) 7 x 2 = ____

12) 2 x 6 = ____

13) 9 x 5 = ____

14) 10 x 7 = ____

15) 10 x 3 = ____

16) 9 x 3 = ____

17) 10 x 9 = ____

18) 4 x 6 = ____

19) 11 x 3 = ____

20) 10 x 4 = ____

Name ________________________ Date ________________

Multiplication Facts

Multiply.

1) 2 x 4	2) 3 x 0	3) 8 x 4	4) 11 x 5	5) 12 x 11
6) 2 x 5	7) 9 x 3	8) 12 x 1	9) 10 x 2	10) 5 x 8
11) 9 x 2	12) 4 x 2	13) 10 x 5	14) 4 x 3	15) 11 x 8
16) 3 x 6	17) 6 x 9	18) 3 x 2	19) 10 x 8	20) 12 x 6

Name ______________________ Date ______________

Multiplication Facts

Multiply.

1) 8 x 3	2) 5 x 5	3) 6 x 3	4) 3 x 0	5) 9 x 6
6) 8 x 1	7) 5 x 8	8) 9 x 7	9) 4 x 9	10) 11 x 7
11) 3 x 3	12) 10 x 5	13) 8 x 7	14) 11 x 6	15) 4 x 6
16) 12 x 6	17) 4 x 2	18) 10 x 3	19) 3 x 6	20) 11 x 5

SECTION

2

FIND THE MISSING MULTIPLIERS

MULTIPLICATION FACTS

11 worksheets
20 problems per sheet

Name ______________________ Date ______________

Multiplication Find the Missing Multipliers

Fill in the blanks.

1) 7x _____ =21

2) 2x _____ =24

3) 9x _____ =18

4) 10x _____ =20

5) 6x _____ =30

6) 3x _____ =15

7) 4x _____ =12

8) 11x _____ =33

9) 8x _____ =80

10) 10x _____ =90

11) 2x _____ =20

12) 4x _____ =20

13) 10x _____ =0

14) 11x _____ =44

15) 4x _____ =44

16) 7x _____ =56

17) 6x _____ =24

18) 10x _____ =80

19) 11x _____ =99

20) 11x _____ =22

Name ______________________________ Date ____________________

Multiplication Find the Missing Multipliers

Fill in the blanks.

1) 12 x _____ = 48

2) 8 x _____ = 48

3) 4 x _____ = 28

4) 8 x _____ = 80

5) 9 x _____ = 45

6) 11 x _____ = 55

7) 3 x _____ = 15

8) 4 x _____ = 16

9) 6 x _____ = 30

10) 8 x _____ = 32

11) 2 x _____ = 0

12) 8 x _____ = 96

13) 2 x _____ = 10

14) 12 x _____ = 36

15) 2 x _____ = 2

16) 2 x _____ = 4

17) 11 x _____ = 77

18) 2 x _____ = 16

19) 12 x _____ = 108

20) 7 x _____ = 63

Name ______________________ Date ______________

Multiplication Find the Missing Multipliers

Fill in the blanks.

1) 10x ____ =110

2) 10x ____ =70

3) 3x ____ =18

4) 10x ____ =30

5) 4x ____ =48

6) 7x ____ =21

7) 11x ____ =22

8) 12x ____ =72

9) 2x ____ =22

10) 8x ____ =40

11) 3x ____ =12

12) 5x ____ =60

13) 6x ____ =30

14) 3x ____ =6

15) 9x ____ =108

16) 10x ____ =10

17) 9x ____ =54

18) 10x ____ =90

19) 8x ____ =0

20) 8x ____ =88

Name ______________________ Date ______________

Multiplication Find the Missing Multipliers

Fill in the blanks.

1) 8 x _____ = 32

2) 3 x _____ = 0

3) 4 x _____ = 48

4) 4 x _____ = 12

5) 5 x _____ = 40

6) 11 x _____ = 121

7) 7 x _____ = 49

8) 8 x _____ = 64

9) 10 x _____ = 60

10) 11 x _____ = 11

11) 2 x _____ = 16

12) 2 x _____ = 4

13) 6 x _____ = 66

14) 4 x _____ = 28

15) 8 x _____ = 96

16) 5 x _____ = 60

17) 11 x _____ = 99

18) 5 x _____ = 55

19) 12 x _____ = 60

20) 7 x _____ = 70

Name ______________________ Date ______________

Multiplication Find the Missing Multipliers

Fill in the blanks.

1) 8x ____ =88

2) 10x ____ =110

3) 2x ____ =0

4) 6x ____ =48

5) 3x ____ =6

6) 9x ____ =54

7) 2x ____ =12

8) 4x ____ =20

9) 10x ____ =70

10) 2x ____ =18

11) 11x ____ =44

12) 10x ____ =40

13) 9x ____ =81

14) 12x ____ =132

15) 2x ____ =4

16) 2x ____ =2

17) 4x ____ =24

18) 3x ____ =12

19) 10x ____ =50

20) 4x ____ =28

Name ______________________ Date ______________________

Multiplication Find the Missing Multipliers

Fill in the blanks.

1) 3 x _____ = 12

2) 6 x _____ = 48

3) 3 x _____ = 9

4) 2 x _____ = 0

5) 5 x _____ = 35

6) 5 x _____ = 50

7) 9 x _____ = 99

8) 8 x _____ = 88

9) 11 x _____ = 99

10) 6 x _____ = 30

11) 3 x _____ = 3

12) 11 x _____ = 121

13) 2 x _____ = 24

14) 10 x _____ = 80

15) 9 x _____ = 36

16) 8 x _____ = 72

17) 12 x _____ = 108

18) 10 x _____ = 120

19) 9 x _____ = 45

20) 7 x _____ = 42

Name ______________________________ Date ____________________

Multiplication Find the Missing Multipliers

Fill in the blanks.

1) 6 x _____ = 6

2) 5 x _____ = 35

3) 6 x _____ = 54

4) 2 x _____ = 6

5) 4 x _____ = 0

6) 8 x _____ = 72

7) 3 x _____ = 30

8) 3 x _____ = 12

9) 2 x _____ = 24

10) 2 x _____ = 14

11) 12 x _____ = 96

12) 4 x _____ = 8

13) 10 x _____ = 60

14) 12 x _____ = 72

15) 8 x _____ = 96

16) 7 x _____ = 28

17) 5 x _____ = 60

18) 4 x _____ = 36

19) 6 x _____ = 66

20) 5 x _____ = 20

Name ____________________ Date ____________________

Multiplication Find the Missing Multipliers

Fill in the blanks.

1) 10 x _____ = 10

2) 4 x _____ = 36

3) 8 x _____ = 40

4) 4 x _____ = 20

5) 4 x _____ = 12

6) 12 x _____ = 132

7) 11 x _____ = 88

8) 7 x _____ = 49

9) 11 x _____ = 55

10) 3 x _____ = 33

11) 9 x _____ = 99

12) 2 x _____ = 0

13) 3 x _____ = 9

14) 2 x _____ = 8

15) 8 x _____ = 56

16) 7 x _____ = 35

17) 4 x _____ = 16

18) 3 x _____ = 18

19) 10 x _____ = 110

20) 9 x _____ = 27

Name ______________________ Date ______________

Multiplication Find the Missing Multipliers

Fill in the blanks.

1) 7x _____ =77

2) 6x _____ =42

3) 10x _____ =70

4) 3x _____ =36

5) 6x _____ =24

6) 2x _____ =18

7) 3x _____ =30

8) 12x _____ =96

9) 9x _____ =99

10) 9x _____ =54

11) 12x _____ =84

12) 10x _____ =80

13) 12x _____ =144

14) 8x _____ =16

15) 7x _____ =70

16) 11x _____ =77

17) 5x _____ =55

18) 4x _____ =24

19) 4x _____ =44

20) 8x _____ =80

Name ______________________________ Date ____________________

Multiplication Find the Missing Multipliers

Fill in the blanks.

1) 11x _____ =22

2) 11x _____ =132

3) 7x _____ =42

4) 2x _____ =10

5) 10x _____ =30

6) 10x _____ =20

7) 11x _____ =110

8) 12x _____ =12

9) 7x _____ =28

10) 7x _____ =0

11) 12x _____ =96

12) 8x _____ =16

13) 4x _____ =40

14) 4x _____ =28

15) 9x _____ =99

16) 6x _____ =48

17) 4x _____ =16

18) 10x _____ =90

19) 6x _____ =42

20) 6x _____ =18

Name ______________________________ Date ____________________

Multiplication Find the Missing Multipliers

Fill in the blanks.

1) 2 x _____ = 16

2) 2 x _____ = 0

3) 12 x _____ = 36

4) 6 x _____ = 54

5) 9 x _____ = 27

6) 3 x _____ = 9

7) 2 x _____ = 4

8) 5 x _____ = 10

9) 2 x _____ = 12

10) 6 x _____ = 12

11) 3 x _____ = 27

12) 4 x _____ = 20

13) 9 x _____ = 108

14) 11 x _____ = 55

15) 7 x _____ = 7

16) 5 x _____ = 60

17) 9 x _____ = 36

18) 10 x _____ = 50

19) 3 x _____ = 36

20) 9 x _____ = 45

SECTION

3

DIVISION FACTS

11 worksheets
20 problems per sheet

Name ____________________ Date ____________________

Division Facts

Divide.

1) $63 \div 7 =$ ______

2) $0 \div 5 =$ ______

3) $120 \div 10 =$ ______

4) $48 \div 8 =$ ______

5) $6 \div 3 =$ ______

6) $66 \div 6 =$ ______

7) $18 \div 9 =$ ______

8) $60 \div 10 =$ ______

9) $7 \div 1 =$ ______

10) $24 \div 6 =$ ______

11) $30 \div 5 =$ ______

12) $72 \div 6 =$ ______

13) $22 \div 11 =$ ______

14) $22 \div 2 =$ ______

15) $12 \div 6 =$ ______

16) $40 \div 4 =$ ______

17) $99 \div 9 =$ ______

18) $80 \div 10 =$ ______

19) $36 \div 12 =$ ______

20) $49 \div 7 =$ ______

Name ______________________ Date ______________

Division Facts

Divide.

1) $64 \div 8 =$ ______

2) $72 \div 6 =$ ______

3) $0 \div 6 =$ ______

4) $36 \div 12 =$ ______

5) $88 \div 8 =$ ______

6) $24 \div 8 =$ ______

7) $9 \div 9 =$ ______

8) $84 \div 7 =$ ______

9) $20 \div 5 =$ ______

10) $2 \div 1 =$ ______

11) $50 \div 5 =$ ______

12) $48 \div 4 =$ ______

13) $12 \div 6 =$ ______

14) $84 \div 12 =$ ______

15) $96 \div 12 =$ ______

16) $24 \div 6 =$ ______

17) $30 \div 6 =$ ______

18) $24 \div 12 =$ ______

19) $18 \div 3 =$ ______

20) $18 \div 9 =$ ______

Name ________________________ Date ________________

Division Facts

Divide.

1) $0 \div 4 =$ ______

2) $50 \div 10 =$ ______

3) $8 \div 4 =$ ______

4) $18 \div 9 =$ ______

5) $44 \div 4 =$ ______

6) $24 \div 6 =$ ______

7) $21 \div 3 =$ ______

8) $63 \div 7 =$ ______

9) $63 \div 9 =$ ______

10) $110 \div 10 =$ ______

11) $90 \div 9 =$ ______

12) $15 \div 3 =$ ______

13) $11 \div 1 =$ ______

14) $108 \div 9 =$ ______

15) $84 \div 12 =$ ______

16) $60 \div 12 =$ ______

17) $12 \div 3 =$ ______

18) $15 \div 5 =$ ______

19) $42 \div 6 =$ ______

20) $120 \div 10 =$ ______

Name ______________________ Date ______________

Division Facts

Divide.

1) $30 \div 3 =$ ______

2) $0 \div 2 =$ ______

3) $27 \div 9 =$ ______

4) $27 \div 3 =$ ______

5) $2 \div 2 =$ ______

6) $42 \div 6 =$ ______

7) $144 \div 12 =$ ______

8) $9 \div 1 =$ ______

9) $77 \div 7 =$ ______

10) $42 \div 7 =$ ______

11) $18 \div 6 =$ ______

12) $12 \div 2 =$ ______

13) $132 \div 11 =$ ______

14) $48 \div 8 =$ ______

15) $49 \div 7 =$ ______

16) $24 \div 8 =$ ______

17) $40 \div 8 =$ ______

18) $16 \div 2 =$ ______

19) $66 \div 6 =$ ______

20) $44 \div 4 =$ ______

Name ______________________ Date ______________

Division Facts

Divide.

1) $56 \div 7 =$ ______

2) $40 \div 4 =$ ______

3) $132 \div 12 =$ ______

4) $72 \div 12 =$ ______

5) $90 \div 10 =$ ______

6) $24 \div 12 =$ ______

7) $45 \div 5 =$ ______

8) $72 \div 6 =$ ______

9) $12 \div 3 =$ ______

10) $8 \div 4 =$ ______

11) $30 \div 3 =$ ______

12) $50 \div 5 =$ ______

13) $10 \div 1 =$ ______

14) $7 \div 7 =$ ______

15) $16 \div 2 =$ ______

16) $24 \div 8 =$ ______

17) $48 \div 6 =$ ______

18) $14 \div 7 =$ ______

19) $15 \div 5 =$ ______

20) $12 \div 6 =$ ______

Name ________________________ Date ________________

Division Facts

Divide.

1) $30 \div 3 =$ ______

2) $44 \div 4 =$ ______

3) $0 \div 12 =$ ______

4) $12 \div 3 =$ ______

5) $72 \div 12 =$ ______

6) $24 \div 6 =$ ______

7) $6 \div 2 =$ ______

8) $32 \div 4 =$ ______

9) $49 \div 7 =$ ______

10) $77 \div 7 =$ ______

11) $9 \div 1 =$ ______

12) $16 \div 4 =$ ______

13) $60 \div 5 =$ ______

14) $2 \div 2 =$ ______

15) $25 \div 5 =$ ______

16) $36 \div 4 =$ ______

17) $28 \div 7 =$ ______

18) $80 \div 10 =$ ______

19) $20 \div 10 =$ ______

20) $42 \div 7 =$ ______

Name ________________________ Date ________________

Division Facts

Divide.

1) $48 \div 4 =$ ______

2) $132 \div 12 =$ ______

3) $132 \div 11 =$ ______

4) $36 \div 4 =$ ______

5) $18 \div 2 =$ ______

6) $3 \div 1 =$ ______

7) $6 \div 6 =$ ______

8) $63 \div 7 =$ ______

9) $108 \div 12 =$ ______

10) $42 \div 6 =$ ______

11) $0 \div 5 =$ ______

12) $40 \div 10 =$ ______

13) $10 \div 5 =$ ______

14) $3 \div 3 =$ ______

15) $84 \div 12 =$ ______

16) $18 \div 6 =$ ______

17) $80 \div 8 =$ ______

18) $30 \div 10 =$ ______

19) $48 \div 8 =$ ______

20) $49 \div 7 =$ ______

Name ______________________ Date ______________

Division Facts

Divide.

1) 9 ÷ 1 = ______

2) 21 ÷ 7 = ______

3) 40 ÷ 10 = ______

4) 24 ÷ 12 = ______

5) 0 ÷ 5 = ______

6) 2 ÷ 2 = ______

7) 63 ÷ 7 = ______

8) 24 ÷ 8 = ______

9) 10 ÷ 10 = ______

10) 20 ÷ 5 = ______

11) 18 ÷ 6 = ______

12) 77 ÷ 7 = ______

13) 32 ÷ 8 = ______

14) 132 ÷ 12 = ______

15) 28 ÷ 7 = ______

16) 54 ÷ 6 = ______

17) 60 ÷ 6 = ______

18) 48 ÷ 8 = ______

19) 30 ÷ 10 = ______

20) 40 ÷ 8 = ______

Name ______________________ Date ______________

Division Facts

Divide.

1) $70 \div 10 =$ ______

2) $35 \div 7 =$ ______

3) $80 \div 8 =$ ______

4) $54 \div 9 =$ ______

5) $84 \div 12 =$ ______

6) $100 \div 10 =$ ______

7) $3 \div 1 =$ ______

8) $0 \div 3 =$ ______

9) $54 \div 6 =$ ______

10) $22 \div 11 =$ ______

11) $99 \div 9 =$ ______

12) $108 \div 9 =$ ______

13) $120 \div 10 =$ ______

14) $10 \div 2 =$ ______

15) $16 \div 8 =$ ______

16) $30 \div 5 =$ ______

17) $8 \div 4 =$ ______

18) $24 \div 8 =$ ______

19) $9 \div 9 =$ ______

20) $16 \div 2 =$ ______

Name ______________________ Date ______________________

Division Facts

Divide.

1) $110 \div 11 =$ ______

2) $35 \div 5 =$ ______

3) $6 \div 2 =$ ______

4) $44 \div 11 =$ ______

5) $84 \div 12 =$ ______

6) $0 \div 6 =$ ______

7) $96 \div 12 =$ ______

8) $36 \div 3 =$ ______

9) $84 \div 7 =$ ______

10) $48 \div 4 =$ ______

11) $16 \div 2 =$ ______

12) $1 \div 1 =$ ______

13) $63 \div 9 =$ ______

14) $24 \div 3 =$ ______

15) $16 \div 4 =$ ______

16) $18 \div 2 =$ ______

17) $9 \div 3 =$ ______

18) $15 \div 5 =$ ______

19) $10 \div 10 =$ ______

20) $99 \div 11 =$ ______

Name ______________________ Date ______________

Division Facts

Divide.

1) $28 \div 4 =$ ______

2) $9 \div 1 =$ ______

3) $22 \div 11 =$ ______

4) $48 \div 4 =$ ______

5) $10 \div 5 =$ ______

6) $0 \div 10 =$ ______

7) $40 \div 4 =$ ______

8) $14 \div 7 =$ ______

9) $60 \div 10 =$ ______

10) $63 \div 7 =$ ______

11) $56 \div 7 =$ ______

12) $144 \div 12 =$ ______

13) $48 \div 6 =$ ______

14) $33 \div 11 =$ ______

15) $90 \div 9 =$ ______

16) $20 \div 4 =$ ______

17) $16 \div 2 =$ ______

18) $6 \div 6 =$ ______

19) $44 \div 11 =$ ______

20) $14 \div 2 =$ ______

SECTION

4

FIND THE MISSING DIVISORS

DIVISION FACTS

11 worksheets
20 problems per sheet

Name ______________________________ Date ____________________

Division Find the Missing Divisors

Fill in the blanks.

1) $3 \div$ ______ $= 1$

2) $1 \div$ ______ $= 1$

3) $30 \div$ ______ $= 3$

4) $18 \div$ ______ $= 6$

5) $12 \div$ ______ $= 3$

6) $54 \div$ ______ $= 6$

7) $48 \div$ ______ $= 8$

8) $20 \div$ ______ $= 10$

9) $60 \div$ ______ $= 5$

10) $24 \div$ ______ $= 4$

11) $28 \div$ ______ $= 7$

12) $40 \div$ ______ $= 8$

13) $12 \div$ ______ $= 6$

14) $80 \div$ ______ $= 8$

15) $16 \div$ ______ $= 4$

16) $48 \div$ ______ $= 4$

17) $88 \div$ ______ $= 11$

18) $56 \div$ ______ $= 8$

19) $90 \div$ ______ $= 10$

20) $15 \div$ ______ $= 3$

Name ______________________ Date ______________

Division Find the Missing Divisors

Fill in the blanks.

1) $108 \div$ _____ $= 9$

2) $18 \div$ _____ $= 2$

3) $96 \div$ _____ $= 8$

4) $11 \div$ _____ $= 11$

5) $16 \div$ _____ $= 4$

6) $20 \div$ _____ $= 4$

7) $32 \div$ _____ $= 8$

8) $18 \div$ _____ $= 9$

9) $49 \div$ _____ $= 7$

10) $45 \div$ _____ $= 5$

11) $24 \div$ _____ $= 8$

12) $10 \div$ _____ $= 1$

13) $121 \div$ _____ $= 11$

14) $63 \div$ _____ $= 9$

15) $84 \div$ _____ $= 12$

16) $16 \div$ _____ $= 8$

17) $12 \div$ _____ $= 1$

18) $33 \div$ _____ $= 3$

19) $33 \div$ _____ $= 11$

20) $100 \div$ _____ $= 10$

Name ______________________ Date ______________

Division Find the Missing Divisors

Fill in the blanks.

1) 50÷ _____ =5

2) 5÷ _____ =1

3) 16÷ _____ =2

4) 70÷ _____ =10

5) 66÷ _____ =11

6) 14÷ _____ =7

7) 84÷ _____ =7

8) 120÷ _____ =12

9) 108÷ _____ =9

10) 99÷ _____ =11

11) 40÷ _____ =8

12) 20÷ _____ =2

13) 40÷ _____ =10

14) 25÷ _____ =5

15) 16÷ _____ =8

16) 12÷ _____ =4

17) 96÷ _____ =12

18) 7÷ _____ =7

19) 49÷ _____ =7

20) 4÷ _____ =2

Name ______________________ Date ______________

Division Find the Missing Divisors

Fill in the blanks.

1) $8 \div$ _____ $= 1$

2) $22 \div$ _____ $= 2$

3) $35 \div$ _____ $= 5$

4) $48 \div$ _____ $= 12$

5) $6 \div$ _____ $= 3$

6) $18 \div$ _____ $= 2$

7) $56 \div$ _____ $= 7$

8) $1 \div$ _____ $= 1$

9) $63 \div$ _____ $= 9$

10) $28 \div$ _____ $= 4$

11) $55 \div$ _____ $= 11$

12) $32 \div$ _____ $= 8$

13) $100 \div$ _____ $= 10$

14) $36 \div$ _____ $= 12$

15) $24 \div$ _____ $= 6$

16) $77 \div$ _____ $= 11$

17) $72 \div$ _____ $= 6$

18) $16 \div$ _____ $= 4$

19) $72 \div$ _____ $= 8$

20) $5 \div$ _____ $= 1$

Name ______________________ Date ______________

Division Find the Missing Divisors

Fill in the blanks.

1) 120÷ _____ =12

2) 45÷ _____ =5

3) 20÷ _____ =10

4) 60÷ _____ =12

5) 88÷ _____ =11

6) 25÷ _____ =5

7) 12÷ _____ =1

8) 36÷ _____ =12

9) 63÷ _____ =7

10) 99÷ _____ =11

11) 54÷ _____ =6

12) 4÷ _____ =4

13) 80÷ _____ =8

14) 5÷ _____ =1

15) 44÷ _____ =11

16) 14÷ _____ =7

17) 42÷ _____ =7

18) 77÷ _____ =11

19) 16÷ _____ =4

20) 24÷ _____ =4

Name ______________________ Date ______________

Division Find the Missing Divisors

Fill in the blanks.

1) 42 ÷ _____ = 7

2) 84 ÷ _____ = 7

3) 3 ÷ _____ = 3

4) 6 ÷ _____ = 3

5) 20 ÷ _____ = 2

6) 33 ÷ _____ = 3

7) 45 ÷ _____ = 9

8) 110 ÷ _____ = 10

9) 16 ÷ _____ = 2

10) 49 ÷ _____ = 7

11) 24 ÷ _____ = 3

12) 40 ÷ _____ = 4

13) 35 ÷ _____ = 5

14) 120 ÷ _____ = 10

15) 108 ÷ _____ = 9

16) 36 ÷ _____ = 3

17) 24 ÷ _____ = 8

18) 12 ÷ _____ = 1

19) 36 ÷ _____ = 9

20) 36 ÷ _____ = 6

Name ______________________ Date ______________

Division Find the Missing Divisors

Fill in the blanks.

1) 44 ÷ _____ = 4

2) 20 ÷ _____ = 4

3) 132 ÷ _____ = 12

4) 10 ÷ _____ = 5

5) 44 ÷ _____ = 11

6) 70 ÷ _____ = 7

7) 100 ÷ _____ = 10

8) 88 ÷ _____ = 8

9) 18 ÷ _____ = 6

10) 16 ÷ _____ = 4

11) 14 ÷ _____ = 7

12) 81 ÷ _____ = 9

13) 12 ÷ _____ = 1

14) 30 ÷ _____ = 6

15) 108 ÷ _____ = 12

16) 24 ÷ _____ = 6

17) 2 ÷ _____ = 2

18) 24 ÷ _____ = 8

19) 24 ÷ _____ = 3

20) 32 ÷ _____ = 8

Name ________________________ Date ________________________

Division Find the Missing Divisors

Fill in the blanks.

1) 55 ÷ _____ = 11

2) 10 ÷ _____ = 1

3) 11 ÷ _____ = 1

4) 10 ÷ _____ = 10

5) 56 ÷ _____ = 7

6) 66 ÷ _____ = 6

7) 24 ÷ _____ = 3

8) 16 ÷ _____ = 8

9) 84 ÷ _____ = 12

10) 25 ÷ _____ = 5

11) 45 ÷ _____ = 5

12) 108 ÷ _____ = 12

13) 77 ÷ _____ = 7

14) 20 ÷ _____ = 5

15) 32 ÷ _____ = 8

16) 63 ÷ _____ = 9

17) 12 ÷ _____ = 4

18) 42 ÷ _____ = 6

19) 72 ÷ _____ = 12

20) 36 ÷ _____ = 12

Name ______________________ Date ______________

Division Find the Missing Divisors

Fill in the blanks.

1) 40÷ _____ =10

2) 45÷ _____ =5

3) 55÷ _____ =5

4) 48÷ _____ =12

5) 12÷ _____ =6

6) 132÷ _____ =12

7) 100÷ _____ =10

8) 4÷ _____ =1

9) 14÷ _____ =2

10) 3÷ _____ =1

11) 42÷ _____ =7

12) 63÷ _____ =7

13) 10÷ _____ =1

14) 11÷ _____ =11

15) 9÷ _____ =3

16) 90÷ _____ =10

17) 60÷ _____ =10

18) 9÷ _____ =1

19) 32÷ _____ =8

20) 32÷ _____ =4

Name ______________________ Date ______________

Division Find the Missing Divisors

Fill in the blanks.

1) $12 \div$ _____ $= 3$

2) $40 \div$ _____ $= 8$

3) $72 \div$ _____ $= 9$

4) $49 \div$ _____ $= 7$

5) $35 \div$ _____ $= 7$

6) $56 \div$ _____ $= 8$

7) $16 \div$ _____ $= 4$

8) $7 \div$ _____ $= 7$

9) $30 \div$ _____ $= 5$

10) $90 \div$ _____ $= 9$

11) $24 \div$ _____ $= 4$

12) $15 \div$ _____ $= 5$

13) $81 \div$ _____ $= 9$

14) $48 \div$ _____ $= 12$

15) $60 \div$ _____ $= 6$

16) $84 \div$ _____ $= 12$

17) $36 \div$ _____ $= 6$

18) $110 \div$ _____ $= 11$

19) $20 \div$ _____ $= 2$

20) $66 \div$ _____ $= 11$

Name ______________________ Date ______________

Division Find the Missing Divisors

Fill in the blanks.

1) 70 ÷ _____ = 10

2) 48 ÷ _____ = 4

3) 54 ÷ _____ = 9

4) 24 ÷ _____ = 6

5) 72 ÷ _____ = 6

6) 28 ÷ _____ = 4

7) 99 ÷ _____ = 11

8) 48 ÷ _____ = 8

9) 35 ÷ _____ = 7

10) 96 ÷ _____ = 12

11) 14 ÷ _____ = 2

12) 1 ÷ _____ = 1

13) 56 ÷ _____ = 8

14) 24 ÷ _____ = 2

15) 21 ÷ _____ = 3

16) 6 ÷ _____ = 2

17) 24 ÷ _____ = 8

18) 66 ÷ _____ = 11

19) 40 ÷ _____ = 4

20) 84 ÷ _____ = 7

SECTION

5

MULTIPLICATION

2 Digits x 1 Digit

11 worksheets
20 problems per sheet

Multiplication — 2-Digit Multiplicands x 1-Digit Multipliers

Multiply.

1)
$$\begin{array}{r} \overset{3}{8}4 \\ \times 9 \\ \hline 756 \end{array}$$

2)
$$\begin{array}{r} \overset{2}{9}7 \\ \times 3 \\ \hline 291 \end{array}$$

3)
$$\begin{array}{r} 11 \\ \times 2 \\ \hline 22 \end{array}$$

4)
$$\begin{array}{r} \overset{4}{5}8 \\ \times 6 \\ \hline 318 \end{array}$$

5)
$$\begin{array}{r} \overset{3}{2}6 \\ \times 6 \\ \hline 156 \end{array}$$

6)
$$\begin{array}{r} 50 \\ \times 3 \\ \hline 250 \end{array}$$

7)
$$\begin{array}{r} \overset{3}{4}7 \\ \times 5 \\ \hline 235 \end{array}$$

8)
$$\begin{array}{r} \overset{2}{5}8 \\ \times 4 \\ \hline 222 \end{array}$$

9)
$$\begin{array}{r} \overset{4}{3}8 \\ \times 6 \\ \hline 218 \end{array}$$

10)
$$\begin{array}{r} 22 \\ \times 4 \\ \hline 88 \end{array}$$

11)
$$\begin{array}{r} 97 \\ \times 1 \\ \hline 97 \end{array}$$

12)
$$\begin{array}{r} 20 \\ \times 2 \\ \hline 40 \end{array}$$

13)
$$\begin{array}{r} \overset{4}{9}7 \\ \times 6 \\ \hline 582 \end{array}$$

14)
$$\begin{array}{r} 81 \\ \times 2 \\ \hline 142 \end{array}$$

15)
$$\begin{array}{r} 41 \\ \times 5 \\ \hline 205 \end{array}$$

16)
$$\begin{array}{r} 91 \\ \times 4 \\ \hline 364 \end{array}$$

17)
$$\begin{array}{r} \overset{2}{2}4 \\ \times 5 \\ \hline 270 \end{array}$$

18)
$$\begin{array}{r} \overset{3}{5}6 \\ \times 6 \\ \hline 336 \end{array}$$

19)
$$\begin{array}{r} 32 \\ \times 2 \\ \hline 64 \end{array}$$

20)
$$\begin{array}{r} \overset{4}{8}6 \\ \times 8 \\ \hline 688 \end{array}$$

Name ____________________ Date ____________________

Multiplication 2-Digit Multiplicands x 1-Digit Multipliers

Multiply.

1) 74 x2	2) 40 x6	3) 67 x1	4) 81 x7	5) 17 x6
6) 25 x9	7) 75 x6	8) 52 x7	9) 46 x8	10) 19 x2
11) 55 x6	12) 44 x5	13) 26 x6	14) 87 x5	15) 39 x5
16) 38 x9	17) 24 x9	18) 60 x9	19) 35 x7	20) 96 x9

Name ______________________ Date ______________

Multiplication 2-Digit Multiplicands x 1-Digit Multipliers

Multiply.

1) 35 x4	2) 52 x4	3) 14 x9	4) 21 x3	5) 49 x3
6) 16 x4	7) 15 x2	8) 56 x2	9) 99 x7	10) 48 x5
11) 86 x6	12) 20 x9	13) 73 x3	14) 47 x5	15) 24 x2
16) 70 x6	17) 37 x7	18) 76 x2	19) 76 x6	20) 95 x7

Name ____________________ Date ____________________

Multiplication 2-Digit Multiplicands x 1-Digit Multipliers

Multiply.

1) 29 x8	2) 94 x8	3) 39 x3	4) 19 x6	5) 66 x8
6) 61 x5	7) 74 x2	8) 72 x8	9) 72 x6	10) 36 x5
11) 22 x7	12) 76 x2	13) 24 x7	14) 86 x7	15) 40 x5
16) 98 x5	17) 79 x1	18) 99 x7	19) 14 x3	20) 28 x7

Name ______________________ Date ______________________

Multiplication 2-Digit Multiplicands x 1-Digit Multipliers

Multiply.

1) 64 x 5

2) 83 x 6

3) 49 x 7

4) 87 x 3

5) 32 x 4

6) 34 x 7

7) 97 x 2

8) 44 x 5

9) 41 x 3

10) 15 x 5

11) 79 x 4

12) 66 x 2

13) 39 x 7

14) 93 x 5

15) 64 x 2

16) 39 x 1

17) 72 x 7

18) 67 x 7

19) 28 x 2

20) 31 x 9

Name ______________________ Date ______________

Multiplication 2-Digit Multiplicands x 1-Digit Multipliers

Multiply.

1) 73 x7	2) 39 x2	3) 67 x6	4) 66 x5	5) 90 x2
6) 42 x4	7) 14 x2	8) 86 x5	9) 30 x8	10) 97 x6
11) 99 x9	12) 13 x2	13) 50 x5	14) 92 x3	15) 28 x4
16) 94 x4	17) 39 x7	18) 95 x4	19) 49 x4	20) 36 x8

Name ______________________ Date ______________

Multiplication 2-Digit Multiplicands x 1-Digit Multipliers

Multiply.

1)	2)	3)	4)	5)
54 x 8	24 x 9	52 x 3	79 x 6	77 x 6

6)	7)	8)	9)	10)
26 x 1	71 x 5	51 x 5	18 x 8	30 x 6

11)	12)	13)	14)	15)
59 x 7	13 x 5	30 x 5	81 x 2	25 x 3

16)	17)	18)	19)	20)
44 x 5	16 x 7	93 x 6	16 x 3	65 x 5

Name ______________________ Date ______________

Multiplication 2-Digit Multiplicands x 1-Digit Multipliers

Multiply.

1) 10 x 5	2) 31 x 1	3) 23 x 8	4) 18 x 3	5) 14 x 5
6) 38 x 3	7) 37 x 4	8) 32 x 6	9) 99 x 5	10) 75 x 2
11) 75 x 7	12) 20 x 4	13) 63 x 2	14) 31 x 3	15) 26 x 6
16) 49 x 4	17) 61 x 6	18) 46 x 7	19) 97 x 5	20) 75 x 4

Name ____________________ Date ____________________

Multiplication 2-Digit Multiplicands x 1-Digit Multipliers

Multiply.

1) 19 x 7	2) 98 x 1	3) 67 x 9	4) 77 x 8	5) 80 x 7
6) 35 x 4	7) 33 x 7	8) 63 x 9	9) 76 x 5	10) 56 x 6
11) 84 x 6	12) 46 x 8	13) 63 x 3	14) 18 x 5	15) 69 x 2
16) 93 x 2	17) 18 x 6	18) 77 x 7	19) 56 x 4	20) 91 x 7

Name ____________________ Date ____________________

Multiplication 2-Digit Multiplicands x 1-Digit Multipliers

Multiply.

1) 94 x7	2) 82 x5	3) 87 x2	4) 71 x2	5) 28 x3
6) 52 x4	7) 39 x4	8) 33 x2	9) 16 x6	10) 68 x3
11) 78 x4	12) 76 x9	13) 33 x6	14) 20 x1	15) 91 x8
16) 88 x5	17) 59 x9	18) 63 x3	19) 56 x5	20) 45 x5

Name ________________________ Date ____________

Multiplication 2-Digit Multiplicands x 1-Digit Multipliers

Multiply.

1) 19 x9	2) 18 x6	3) 68 x9	4) 10 x3	5) 72 x4
6) 44 x5	7) 61 x7	8) 40 x7	9) 21 x9	10) 51 x7
11) 73 x5	12) 84 x5	13) 22 x3	14) 14 x8	15) 81 x9
16) 29 x2	17) 54 x4	18) 92 x9	19) 63 x9	20) 32 x4

SECTION

6

MULTIPLICATION

4 Digits x 1 Digit

11 worksheets
15 problems per sheet

Name ______________________ Date ______________

Multiplication 4-Digit Multiplicands x 1-Digit Multipliers

Multiply.

1) 2159 x6	2) 1826 x3	3) 4900 x4
4) 3328 x5	5) 6780 x7	6) 4391 x5
7) 7517 x3	8) 2482 x4	9) 2330 x9
10) 8572 x3	11) 8938 x6	12) 2129 x3
13) 4560 x2	14) 3121 x4	15) 6915 x3

Name ______________________ Date ______________

Multiplication 4-Digit Multiplicands x 1-Digit Multipliers

Multiply.

1) 9017 x7

2) 2220 x8

3) 3786 x5

4) 8196 x8

5) 2898 x8

6) 2060 x2

7) 8319 x5

8) 1532 x3

9) 8520 x9

10) 1829 x4

11) 5470 x2

12) 2672 x2

13) 5635 x3

14) 8278 x6

15) 1987 x2

Name ______________________________ Date ____________________

Multiplication 4-Digit Multiplicands x 1-Digit Multipliers

Multiply.

1) 8728
x3

2) 1694
x9

3) 9843
x7

4) 7968
x5

5) 4331
x2

6) 2413
x7

7) 7661
x5

8) 7246
x8

9) 3802
x8

10) 4186
x4

11) 8222
x4

12) 6452
x4

13) 4889
x2

14) 4764
x5

15) 9274
x3

Name ______________________ Date ______________________

Multiplication 4-Digit Multiplicands x 1-Digit Multipliers

Multiply.

1) 3339 x8

2) 5344 x7

3) 1896 x8

4) 8569 x8

5) 4655 x8

6) 2916 x5

7) 2334 x4

8) 3856 x8

9) 8151 x4

10) 4356 x9

11) 3825 x5

12) 1885 x2

13) 2576 x9

14) 7412 x2

15) 4122 x3

Name ______________________ Date ______________

Multiplication 4-Digit Multiplicands x 1-Digit Multipliers

Multiply.

1) 9406 x4	2) 3682 x5	3) 9836 x6
4) 9973 x3	5) 6169 x4	6) 9216 x6
7) 9376 x3	8) 1129 x3	9) 7925 x7
10) 7931 x2	11) 3784 x6	12) 2340 x9
13) 3423 x3	14) 6575 x8	15) 8809 x6

Name ____________________ Date ____________________

Multiplication 4-Digit Multiplicands x 1-Digit Multipliers

Multiply.

1) 2355 x4	2) 8413 x8	3) 2899 x5
4) 4202 x8	5) 7926 x8	6) 2645 x2
7) 3653 x7	8) 5711 x4	9) 7800 x5
10) 1869 x8	11) 6586 x4	12) 2088 x6
13) 8730 x8	14) 3836 x7	15) 9449 x3

Name ________________ Date ________________

Multiplication 4-Digit Multiplicands x 1-Digit Multipliers

Multiply.

1) 2198 x8	2) 8229 x6	3) 4447 x6
4) 3051 x8	5) 6960 x6	6) 7559 x6
7) 9455 x5	8) 1221 x8	9) 7151 x2
10) 4006 x6	11) 8767 x3	12) 4383 x3
13) 9522 x9	14) 4320 x6	15) 9134 x8

Name ______________________ Date ______________________

Multiplication 4-Digit Multiplicands x 1-Digit Multipliers

Multiply.

1) 1287 x 9 = ____

2) 2822 x 9 = ____

3) 6252 x 2 = ____

4) 8306 x 8 = ____

5) 7683 x 5 = ____

6) 2803 x 8 = ____

7) 7550 x 9 = ____

8) 4032 x 3 = ____

9) 3322 x 8 = ____

10) 7213 x 4 = ____

11) 5966 x 8 = ____

12) 3063 x 6 = ____

13) 3952 x 6 = ____

14) 4802 x 3 = ____

15) 8249 x 7 = ____

Name ____________________ Date ____________________

Multiplication 4-Digit Multiplicands x 1-Digit Multipliers

Multiply.

1) 3149 x8	2) 7325 x6	3) 2321 x4
4) 3149 x9	5) 2039 x5	6) 2190 x5
7) 6520 x4	8) 6417 x7	9) 6484 x5
10) 4297 x2	11) 6149 x9	12) 6801 x5
13) 5234 x5	14) 2397 x3	15) 7957 x9

Name ______________________ Date ______________

Multiplication 4-Digit Multiplicands x 1-Digit Multipliers

Multiply.

1) 4910 x7	2) 1893 x5	3) 4285 x2
4) 5443 x2	5) 9689 x7	6) 3069 x8
7) 5092 x2	8) 3489 x6	9) 5961 x6
10) 1924 x2	11) 5265 x3	12) 7091 x4
13) 9474 x4	14) 5209 x2	15) 2556 x6

Name ______________________ Date ______________________

Multiplication 4-Digit Multiplicands x 1-Digit Multipliers

Multiply.

1) 7355 x9	2) 8587 x5	3) 4928 x2
4) 7453 x8	5) 8953 x4	6) 8691 x8
7) 1147 x5	8) 9771 x9	9) 3730 x3
10) 3661 x2	11) 7420 x5	12) 5831 x5
13) 7437 x4	14) 9249 x8	15) 3498 x2

SECTION

7

MULTIPLICATION

3 Digits x 2 Digits

11 worksheets
9 problems per sheet

Name ______________________ Date ______________

Multiplication 3-Digit Multiplicands x 2-Digit Multipliers

Multiply.

1) 120 x79

2) 873 x34

3) 840 x82

4) 833 x75

5) 623 x43

6) 517 x78

7) 229 x48

8) 459 x18

9) 866 x50

Name ______________________ Date ______________

Multiplication 3-Digit Multiplicands x 2-Digit Multipliers

Multiply.

1) 811 x43

2) 837 x58

3) 263 x37

4) 380 x93

5) 761 x91

6) 841 x70

7) 589 x45

8) 282 x74

9) 282 x53

Name ______________________ Date ______________

Multiplication 3-Digit Multiplicands x 2-Digit Multipliers

Multiply.

1) 307 x86	2) 865 x74	3) 145 x10
4) 871 x73	5) 227 x78	6) 432 x85
7) 413 x16	8) 982 x84	9) 112 x47

Name ______________________ Date ______________________

Multiplication 3-Digit Multiplicands x 2-Digit Multipliers

Multiply.

1) 572 x36

2) 879 x74

3) 765 x51

4) 368 x20

5) 583 x34

6) 143 x29

7) 787 x74

8) 739 x76

9) 784 x32

Name ______________________ Date ______________

Multiplication 3-Digit Multiplicands x 2-Digit Multipliers

Multiply.

1) 729 x16	2) 219 x37	3) 951 x32
4) 503 x58	5) 138 x12	6) 976 x59
7) 173 x67	8) 595 x38	9) 692 x81

Name ______________________ Date ______________

Multiplication 3-Digit Multiplicands x 2-Digit Multipliers

Multiply.

1) 770 x29	2) 978 x92	3) 157 x34
4) 892 x31	5) 899 x52	6) 909 x19
7) 831 x56	8) 439 x11	9) 937 x67

Name ______________________ Date ______________

Multiplication 3-Digit Multiplicands x 2-Digit Multipliers

Multiply.

1) 441 x91

2) 436 x85

3) 682 x47

4) 950 x96

5) 647 x56

6) 387 x62

7) 870 x91

8) 856 x25

9) 537 x82

Name ______________________ Date ______________

Multiplication 3-Digit Multiplicands x 2-Digit Multipliers

Multiply.

1) 239 x41

2) 238 x56

3) 225 x45

4) 448 x28

5) 117 x23

6) 775 x23

7) 675 x63

8) 854 x17

9) 409 x27

Name ______________________ Date ______________________

Multiplication 3-Digit Multiplicands x 2-Digit Multipliers

Multiply.

1) 254 x89	2) 828 x52	3) 240 x58
4) 680 x40	5) 823 x41	6) 101 x13
7) 285 x60	8) 751 x60	9) 962 x88

Name ______________________ Date ______________________

Multiplication 3-Digit Multiplicands x 2-Digit Multipliers

Multiply.

1) 931 x88

2) 531 x20

3) 377 x60

4) 777 x34

5) 545 x77

6) 382 x74

7) 845 x90

8) 329 x71

9) 237 x98

Name ______________________ Date ______________

Multiplication 3-Digit Multiplicands x 2-Digit Multipliers

Multiply.

1) 744 x93

2) 595 x47

3) 106 x10

4) 625 x79

5) 430 x57

6) 814 x56

7) 450 x29

8) 572 x18

9) 250 x37

SECTION

8

DIVISION

3 Digits / 1 Digit

11 worksheets
12 problems per sheet

Name ______________________ Date ______________

Division 3-Digit Dividends / 1-Digit Divisors

Divide.

1) $4\overline{)192}$

2) $4\overline{)348}$

3) $5\overline{)115}$

4) $6\overline{)150}$

5) $3\overline{)108}$

6) $9\overline{)522}$

7) $3\overline{)147}$

8) $2\overline{)146}$

9) $9\overline{)144}$

10) $2\overline{)180}$

11) $6\overline{)492}$

12) $7\overline{)497}$

Name ________________________ Date ________________

Division 3-Digit Dividends / 1-Digit Divisors

Divide.

1) $8\overline{)720}$

2) $9\overline{)270}$

3) $5\overline{)340}$

4) $2\overline{)120}$

5) $7\overline{)497}$

6) $8\overline{)664}$

7) $2\overline{)116}$

8) $7\overline{)588}$

9) $8\overline{)752}$

10) $8\overline{)136}$

11) $4\overline{)212}$

12) $2\overline{)156}$

Name ______________________ Date ______________

Division 3-Digit Dividends / 1-Digit Divisors

Divide.

1) $7\overline{)343}$

2) $4\overline{)200}$

3) $4\overline{)196}$

4) $2\overline{)170}$

5) $7\overline{)588}$

6) $4\overline{)160}$

7) $2\overline{)114}$

8) $5\overline{)125}$

9) $6\overline{)198}$

10) $8\overline{)584}$

11) $9\overline{)234}$

12) $9\overline{)756}$

Name ______________________ Date ______________________

Division 3-Digit Dividends / 1-Digit Divisors

Divide.

1) $4\overline{)336}$

2) $6\overline{)528}$

3) $5\overline{)155}$

4) $4\overline{)280}$

5) $4\overline{)216}$

6) $5\overline{)280}$

7) $9\overline{)117}$

8) $5\overline{)270}$

9) $9\overline{)675}$

10) $5\overline{)180}$

11) $9\overline{)234}$

12) $9\overline{)594}$

Name ____________________ Date ____________

Division 3-Digit Dividends / 1-Digit Divisors

Divide.

1) $5\overline{)335}$

2) $5\overline{)345}$

3) $3\overline{)282}$

4) $6\overline{)192}$

5) $3\overline{)120}$

6) $3\overline{)147}$

7) $7\overline{)476}$

8) $4\overline{)192}$

9) $8\overline{)112}$

10) $6\overline{)258}$

11) $8\overline{)696}$

12) $6\overline{)438}$

Name ______________________ Date ______________________

Division 3-Digit Dividends / 1-Digit Divisors

Divide.

1) $3\overline{)168}$

2) $9\overline{)207}$

3) $4\overline{)196}$

4) $8\overline{)728}$

5) $8\overline{)504}$

6) $7\overline{)518}$

7) $4\overline{)212}$

8) $7\overline{)567}$

9) $5\overline{)430}$

10) $9\overline{)711}$

11) $9\overline{)189}$

12) $3\overline{)198}$

Name ______________________ Date ______________

Division 3-Digit Dividends / 1-Digit Divisors

Divide.

1) $8\overline{)288}$

2) $7\overline{)112}$

3) $4\overline{)256}$

4) $8\overline{)640}$

5) $9\overline{)189}$

6) $9\overline{)585}$

7) $6\overline{)138}$

8) $5\overline{)295}$

9) $5\overline{)465}$

10) $8\overline{)352}$

11) $8\overline{)560}$

12) $9\overline{)162}$

Name ______________________ Date ______________________

Division 3-Digit Dividends / 1-Digit Divisors

Divide.

1) $9\overline{)198}$

2) $4\overline{)180}$

3) $9\overline{)486}$

4) $2\overline{)132}$

5) $4\overline{)224}$

6) $9\overline{)342}$

7) $9\overline{)585}$

8) $8\overline{)336}$

9) $7\overline{)637}$

10) $8\overline{)704}$

11) $3\overline{)258}$

12) $9\overline{)432}$

Name ______________________ Date ______________

Division 3-Digit Dividends / 1-Digit Divisors

Divide.

1) $5\overline{)385}$

2) $6\overline{)108}$

3) $8\overline{)120}$

4) $8\overline{)648}$

5) $7\overline{)497}$

6) $4\overline{)364}$

7) $2\overline{)160}$

8) $3\overline{)144}$

9) $3\overline{)165}$

10) $5\overline{)250}$

11) $7\overline{)511}$

12) $8\overline{)232}$

Name ______________________________ Date ____________________

Division 3-Digit Dividends / 1-Digit Divisors

Divide.

1) $5\overline{)240}$

2) $4\overline{)136}$

3) $8\overline{)376}$

4) $8\overline{)392}$

5) $2\overline{)128}$

6) $4\overline{)296}$

7) $7\overline{)525}$

8) $9\overline{)567}$

9) $8\overline{)776}$

10) $8\overline{)104}$

11) $2\overline{)108}$

12) $5\overline{)175}$

Name ______________________ Date ______________

Division 3-Digit Dividends / 1-Digit Divisors

Divide.

1) $7\overline{)665}$

2) $7\overline{)238}$

3) $9\overline{)801}$

4) $8\overline{)272}$

5) $7\overline{)231}$

6) $4\overline{)392}$

7) $3\overline{)171}$

8) $5\overline{)350}$

9) $8\overline{)104}$

10) $7\overline{)175}$

11) $4\overline{)136}$

12) $8\overline{)760}$

SECTION

9

DIVISION

5 Digits / 1 Digit

11 worksheets
9 problems per sheet

Name ______________________ Date ______________

Division 5-Digit Dividends / 1-Digit Divisors

Divide.

1) $4\overline{)86852}$

2) $3\overline{)28554}$

3) $4\overline{)48432}$

4) $3\overline{)31431}$

5) $9\overline{)42687}$

6) $9\overline{)82152}$

7) $2\overline{)29984}$

8) $2\overline{)13410}$

9) $2\overline{)41836}$

Name ______________________ Date ______________________

Division 5-Digit Dividends / 1-Digit Divisors

Divide.

1) $6\overline{)30810}$

2) $4\overline{)15896}$

3) $8\overline{)79648}$

4) $4\overline{)40940}$

5) $5\overline{)60440}$

6) $9\overline{)47961}$

7) $2\overline{)30560}$

8) $9\overline{)43956}$

9) $3\overline{)40830}$

Name ______________________ Date ______________

Division 5-Digit Dividends / 1-Digit Divisors

Divide.

1) $2\overline{)16516}$

2) $9\overline{)40653}$

3) $5\overline{)16980}$

4) $3\overline{)40797}$

5) $2\overline{)28290}$

6) $3\overline{)11970}$

7) $2\overline{)15096}$

8) $4\overline{)74944}$

9) $9\overline{)85455}$

Name ______________________ Date ______________________

Division 5-Digit Dividends / 1-Digit Divisors

Divide.

1) $7\overline{)76552}$

2) $3\overline{)52533}$

3) $7\overline{)92554}$

4) $5\overline{)91550}$

5) $6\overline{)70392}$

6) $3\overline{)29421}$

7) $5\overline{)24000}$

8) $3\overline{)62553}$

9) $6\overline{)46752}$

Name ____________________ Date ____________________

Division 5-Digit Dividends / 1-Digit Divisors

Divide.

1) $2\overline{)39768}$

2) $2\overline{)25988}$

3) $7\overline{)28595}$

4) $9\overline{)38970}$

5) $7\overline{)85260}$

6) $6\overline{)98244}$

7) $3\overline{)40935}$

8) $5\overline{)98340}$

9) $3\overline{)40386}$

Name ______________________ Date ______________________

Division 5-Digit Dividends / 1-Digit Divisors

Divide.

1) $6\overline{)87366}$

2) $4\overline{)63472}$

3) $5\overline{)37205}$

4) $3\overline{)51516}$

5) $4\overline{)57732}$

6) $9\overline{)94059}$

7) $3\overline{)34860}$

8) $2\overline{)30864}$

9) $3\overline{)10437}$

Name ______________________ Date ______________

Division 5-Digit Dividends / 1-Digit Divisors

Divide.

1) $9\overline{)52308}$

2) $2\overline{)25566}$

3) $6\overline{)81630}$

4) $3\overline{)46341}$

5) $9\overline{)22401}$

6) $5\overline{)74210}$

7) $3\overline{)13548}$

8) $3\overline{)46449}$

9) $3\overline{)41697}$

Name ______________________ Date ______________

Division 5-Digit Dividends / 1-Digit Divisors

Divide.

1) $9\overline{)57933}$

2) $6\overline{)45840}$

3) $3\overline{)24480}$

4) $2\overline{)36022}$

5) $9\overline{)66186}$

6) $5\overline{)97580}$

7) $7\overline{)68082}$

8) $3\overline{)28437}$

9) $8\overline{)27464}$

Name ______________________ Date ______________

Division 5-Digit Dividends / 1-Digit Divisors

Divide.

1) $4\overline{)59324}$

2) $3\overline{)31209}$

3) $2\overline{)12522}$

4) $4\overline{)76988}$

5) $7\overline{)59563}$

6) $8\overline{)97848}$

7) $6\overline{)37296}$

8) $6\overline{)22200}$

9) $7\overline{)75516}$

Name ______________________ Date ______________

Division 5-Digit Dividends / 1-Digit Divisors

Divide.

1) $7\overline{)22645}$

2) $4\overline{)69184}$

3) $7\overline{)79884}$

4) $7\overline{)35805}$

5) $7\overline{)85071}$

6) $4\overline{)80996}$

7) $6\overline{)88296}$

8) $4\overline{)17376}$

9) $4\overline{)17952}$

Name ______________________________ Date ____________________

Division 5-Digit Dividends / 1-Digit Divisors

Divide.

1) $3\overline{)41601}$

2) $4\overline{)82012}$

3) $4\overline{)82888}$

4) $4\overline{)72008}$

5) $4\overline{)38584}$

6) $6\overline{)14916}$

7) $4\overline{)20468}$

8) $6\overline{)86754}$

9) $5\overline{)36895}$

SECTION

10

DIVISION

4 Digits / 1 Digit
with Remainders

11 worksheets
9 problems per sheet

Name ______________________ Date ______________

Division 4-Digit Dividends / 1-Digit Divisors *(remainders)*

Divide.

1) $7\overline{)3191}$

2) $5\overline{)3846}$

3) $6\overline{)7665}$

4) $7\overline{)2774}$

5) $6\overline{)8221}$

6) $2\overline{)9345}$

7) $9\overline{)9038}$

8) $6\overline{)7682}$

9) $7\overline{)9838}$

Name ______________________________ Date ____________________

Division 4-Digit Dividends / 1-Digit Divisors *(remainders)*

Divide.

1) $9\overline{)4190}$ 2) $2\overline{)5885}$ 3) $3\overline{)9734}$

4) $8\overline{)5814}$ 5) $8\overline{)3979}$ 6) $9\overline{)2834}$

7) $6\overline{)6484}$ 8) $4\overline{)3826}$ 9) $3\overline{)5272}$

Name ______________________ Date ______________

Division 4-Digit Dividends / 1-Digit Divisors *(remainders)*

Divide.

1) $6\overline{)7585}$ 2) $8\overline{)9476}$ 3) $2\overline{)9377}$

4) $6\overline{)8924}$ 5) $3\overline{)1813}$ 6) $5\overline{)1488}$

7) $6\overline{)4621}$ 8) $8\overline{)8009}$ 9) $9\overline{)6136}$

Name ____________________ Date ____________________

Division 4-Digit Dividends / 1-Digit Divisors *(remainders)*

Divide.

1) $6\overline{)1363}$

2) $2\overline{)8007}$

3) $5\overline{)7787}$

4) $9\overline{)5836}$

5) $8\overline{)8492}$

6) $6\overline{)8603}$

7) $4\overline{)6550}$

8) $8\overline{)6895}$

9) $9\overline{)7237}$

Name ____________________ Date ____________________

Division 4-Digit Dividends / 1-Digit Divisors *(remainders)*

Divide.

1) $7\overline{)2314}$

2) $2\overline{)2011}$

3) $6\overline{)3340}$

4) $3\overline{)4456}$

5) $5\overline{)7692}$

6) $9\overline{)8416}$

7) $5\overline{)6926}$

8) $8\overline{)2972}$

9) $4\overline{)6394}$

Name ______________________ Date ______________

Division 4-Digit Dividends / 1-Digit Divisors *(remainders)*

Divide.

1) $4\overline{)6714}$

2) $7\overline{)8770}$

3) $5\overline{)3753}$

4) $5\overline{)7182}$

5) $3\overline{)7807}$

6) $7\overline{)3813}$

7) $8\overline{)1662}$

8) $3\overline{)7711}$

9) $8\overline{)3451}$

Name ______________________ Date ______________

Division 4-Digit Dividends / 1-Digit Divisors *(remainders)*

Divide.

1) $3\overline{)6082}$

2) $9\overline{)7907}$

3) $3\overline{)3182}$

4) $6\overline{)3710}$

5) $5\overline{)5826}$

6) $2\overline{)8357}$

7) $5\overline{)1899}$

8) $8\overline{)4501}$

9) $9\overline{)6248}$

Name ______________________ Date ______________

Division 4-Digit Dividends / 1-Digit Divisors *(remainders)*

Divide.

1) $7\overline{)6567}$

2) $6\overline{)8092}$

3) $8\overline{)5892}$

4) $3\overline{)4123}$

5) $6\overline{)2815}$

6) $3\overline{)8507}$

7) $6\overline{)9867}$

8) $8\overline{)3039}$

9) $4\overline{)7907}$

Name ______________________ Date ______________________

Division 4-Digit Dividends / 1-Digit Divisors *(remainders)*

Divide.

1) $9\overline{)8418}$

2) $3\overline{)1573}$

3) $2\overline{)1645}$

4) $6\overline{)2164}$

5) $8\overline{)2958}$

6) $7\overline{)3573}$

7) $5\overline{)5522}$

8) $2\overline{)2329}$

9) $3\overline{)3275}$

Name ______________________ Date ______________

Division 4-Digit Dividends / 1-Digit Divisors *(remainders)*

Divide.

1) $7\overline{)5468}$

2) $5\overline{)8873}$

3) $3\overline{)4667}$

4) $9\overline{)8686}$

5) $9\overline{)2534}$

6) $8\overline{)4422}$

7) $7\overline{)1963}$

8) $5\overline{)5777}$

9) $7\overline{)7115}$

Name ____________________ Date ____________________

Division 4-Digit Dividends / 1-Digit Divisors *(remainders)*

Divide.

1) $5\overline{)5843}$

2) $8\overline{)6782}$

3) $2\overline{)8725}$

4) $9\overline{)5197}$

5) $9\overline{)9722}$

6) $9\overline{)6709}$

7) $8\overline{)6084}$

8) $8\overline{)4793}$

9) $9\overline{)2662}$

SECTION

11

DIVISION

4 Digits / 2 Digits

11 worksheets
9 problems per sheet

Name ______________________ Date ______________

Division 4-Digit Dividends / 2-Digit Divisors

Divide.

1) $51\overline{)4284}$

2) $29\overline{)5626}$

3) $54\overline{)9396}$

4) $10\overline{)9840}$

5) $46\overline{)5336}$

6) $15\overline{)5715}$

7) $21\overline{)5460}$

8) $71\overline{)6887}$

9) $41\overline{)2419}$

Name ____________________ Date ____________________

Division 4-Digit Dividends / 2-Digit Divisors

Divide.

1) $69\overline{)2898}$

2) $35\overline{)5075}$

3) $48\overline{)2208}$

4) $64\overline{)2432}$

5) $89\overline{)1691}$

6) $19\overline{)4579}$

7) $44\overline{)1100}$

8) $28\overline{)3416}$

9) $39\overline{)9906}$

Name ______________________ Date ______________

Division 4-Digit Dividends / 2-Digit Divisors

Divide.

1) $27\overline{)4725}$

2) $21\overline{)3591}$

3) $16\overline{)5824}$

4) $80\overline{)1040}$

5) $40\overline{)4480}$

6) $85\overline{)5950}$

7) $20\overline{)5540}$

8) $33\overline{)8283}$

9) $31\overline{)4960}$

Name ______________________ Date ______________

Division 4-Digit Dividends / 2-Digit Divisors

Divide.

1) $22\overline{)8888}$

2) $12\overline{)6480}$

3) $32\overline{)2272}$

4) $23\overline{)1518}$

5) $21\overline{)2100}$

6) $86\overline{)4558}$

7) $11\overline{)3388}$

8) $26\overline{)6864}$

9) $48\overline{)8208}$

Name ______________________ Date ______________

Division 4-Digit Dividends / 2-Digit Divisors

Divide.

1) $22\overline{)2618}$

2) $13\overline{)9581}$

3) $18\overline{)4014}$

4) $89\overline{)4895}$

5) $80\overline{)8160}$

6) $15\overline{)3855}$

7) $75\overline{)6075}$

8) $16\overline{)7248}$

9) $24\overline{)7152}$

Name ______________________________ Date ____________________

Division 4-Digit Dividends / 2-Digit Divisors

Divide.

1) $88\overline{)9504}$

2) $11\overline{)7898}$

3) $70\overline{)7980}$

4) $10\overline{)7350}$

5) $12\overline{)7548}$

6) $45\overline{)7605}$

7) $28\overline{)8932}$

8) $64\overline{)4736}$

9) $57\overline{)2964}$

Name ____________________ Date ____________________

Division 4-Digit Dividends / 2-Digit Divisors

Divide.

1) $17\overline{)5287}$

2) $40\overline{)3080}$

3) $36\overline{)4608}$

4) $21\overline{)6216}$

5) $62\overline{)1178}$

6) $15\overline{)1440}$

7) $26\overline{)5928}$

8) $44\overline{)2684}$

9) $52\overline{)1612}$

Name ______________________ Date ______________

Division 4-Digit Dividends / 2-Digit Divisors

Divide.

1) $31\overline{)6076}$

2) $77\overline{)4928}$

3) $21\overline{)8841}$

4) $56\overline{)7840}$

5) $37\overline{)8695}$

6) $61\overline{)4392}$

7) $55\overline{)6710}$

8) $40\overline{)9480}$

9) $45\overline{)5985}$

Name ______________________ Date ______________

Division 4-Digit Dividends / 2-Digit Divisors

Divide.

1) $71\overline{)4331}$

2) $12\overline{)2628}$

3) $27\overline{)2322}$

4) $35\overline{)8190}$

5) $81\overline{)3969}$

6) $38\overline{)4218}$

7) $60\overline{)5340}$

8) $12\overline{)6048}$

9) $16\overline{)5248}$

Name ______________________ Date ______________

Division 4-Digit Dividends / 2-Digit Divisors

Divide.

1) $31\overline{)8835}$

2) $20\overline{)1280}$

3) $34\overline{)6732}$

4) $11\overline{)2596}$

5) $10\overline{)3250}$

6) $60\overline{)7500}$

7) $87\overline{)7743}$

8) $79\overline{)5214}$

9) $33\overline{)7095}$

Name ______________________ Date ______________

Division 4-Digit Dividends / 2-Digit Divisors

Divide.

1) $44\overline{)8184}$

2) $18\overline{)5292}$

3) $85\overline{)2890}$

4) $32\overline{)9728}$

5) $22\overline{)8096}$

6) $23\overline{)8073}$

7) $11\overline{)6523}$

8) $32\overline{)3424}$

9) $10\overline{)1640}$

SECTION

12

MULTIPLICATION
by 10, 100, 1000

11 worksheets
20 problems per sheet

Name ______________________ Date ______________

Multiplication Multiply by 10, 100 and 1000

Multiply.

1) 59x100= ________

2) 93x10= ________

3) 92x100= ________

4) 42x10= ________

5) 58x10= ________

6) 41x10= ________

7) 20x1000= ________

8) 68x1000= ________

9) 43x1000= ________

10) 18x1000= ________

11) 22x1000= ________

12) 32x10= ________

13) 69x10= ________

14) 14x1000= ________

15) 44x1000= ________

16) 40x10= ________

17) 95x10= ________

18) 90x100= ________

19) 32x100= ________

20) 57x100= ________

Name ______________________ Date ______________

Multiplication Multiply by 10, 100 and 1000

Multiply.

1) 21x10= ________

2) 38x10= ________

3) 16x100= ________

4) 24x1000= ________

5) 92x100= ________

6) 38x100= ________

7) 60x10= ________

8) 59x100= ________

9) 66x10= ________

10) 67x100= ________

11) 87x10= ________

12) 89x1000= ________

13) 70x1000= ________

14) 81x1000= ________

15) 45x100= ________

16) 96x10= ________

17) 74x1000= ________

18) 28x1000= ________

19) 52x1000= ________

20) 26x100= ________

Name ______________________ Date ______________

Multiplication Multiply by 10, 100 and 1000

Multiply.

1) 90x1000= ________

2) 39x100= ________

3) 26x10= ________

4) 40x10= ________

5) 23x100= ________

6) 48x10= ________

7) 58x100= ________

8) 42x10= ________

9) 27x1000= ________

10) 86x1000= ________

11) 43x1000= ________

12) 69x100= ________

13) 26x100= ________

14) 49x100= ________

15) 63x10= ________

16) 32x1000= ________

17) 49x10= ________

18) 80x100= ________

19) 48x100= ________

20) 88x10= ________

Name ______________________ Date ______________________

Multiplication Multiply by 10, 100 and 1000

Multiply.

1) 77x1000= ________

2) 73x100= ________

3) 18x100= ________

4) 68x100= ________

5) 82x10= ________

6) 49x100= ________

7) 80x1000= ________

8) 79x10= ________

9) 75x10= ________

10) 53x1000= ________

11) 38x100= ________

12) 34x100= ________

13) 84x100= ________

14) 67x1000= ________

15) 78x1000= ________

16) 81x10= ________

17) 85x10= ________

18) 55x100= ________

19) 83x100= ________

20) 51x1000= ________

Name ______________________ Date ______________

Multiplication Multiply by 10, 100 and 1000

Multiply.

1) 44x10= ________

2) 35x100= ________

3) 53x100= ________

4) 87x100= ________

5) 80x100= ________

6) 20x100= ________

7) 19x100= ________

8) 65x1000= ________

9) 18x1000= ________

10) 64x10= ________

11) 51x100= ________

12) 50x100= ________

13) 54x10= ________

14) 29x100= ________

15) 69x1000= ________

16) 53x1000= ________

17) 75x10= ________

18) 15x1000= ________

19) 34x1000= ________

20) 56x100= ________

Name ______________________ Date ______________

Multiplication Multiply by 10, 100 and 1000

Multiply.

1) 81x100= ________

2) 89x1000= ________

3) 16x1000= ________

4) 61x10= ________

5) 18x10= ________

6) 87x10= ________

7) 64x100= ________

8) 41x100= ________

9) 69x1000= ________

10) 71x1000= ________

11) 60x1000= ________

12) 69x100= ________

13) 38x100= ________

14) 42x100= ________

15) 73x1000= ________

16) 55x100= ________

17) 17x10= ________

18) 32x10= ________

19) 44x1000= ________

20) 97x1000= ________

Name ______________________ Date ______________

Multiplication Multiply by 10, 100 and 1000

Multiply.

1) 45x100= ________

2) 52x100= ________

3) 86x100= ________

4) 51x10= ________

5) 60x100= ________

6) 15x1000= ________

7) 31x100= ________

8) 18x100= ________

9) 28x10= ________

10) 14x1000= ________

11) 71x1000= ________

12) 43x100= ________

13) 20x10= ________

14) 54x100= ________

15) 67x100= ________

16) 25x100= ________

17) 93x10= ________

18) 62x1000= ________

19) 93x1000= ________

20) 47x10= ________

Name ______________________ Date ______________

Multiplication Multiply by 10, 100 and 1000

Multiply.

1) $70 \times 1000 =$ ________

2) $76 \times 100 =$ ________

3) $40 \times 100 =$ ________

4) $97 \times 10 =$ ________

5) $86 \times 10 =$ ________

6) $57 \times 1000 =$ ________

7) $17 \times 100 =$ ________

8) $18 \times 1000 =$ ________

9) $67 \times 10 =$ ________

10) $45 \times 10 =$ ________

11) $79 \times 1000 =$ ________

12) $50 \times 1000 =$ ________

13) $30 \times 100 =$ ________

14) $50 \times 10 =$ ________

15) $95 \times 1000 =$ ________

16) $25 \times 1000 =$ ________

17) $81 \times 100 =$ ________

18) $39 \times 1000 =$ ________

19) $54 \times 10 =$ ________

20) $71 \times 1000 =$ ________

Name ____________________ Date ____________________

Multiplication Multiply by 10, 100 and 1000

Multiply.

1) 37x10= ________

2) 97x100= ________

3) 46x10= ________

4) 20x10= ________

5) 25x1000= ________

6) 29x1000= ________

7) 38x10= ________

8) 19x100= ________

9) 64x10= ________

10) 53x10= ________

11) 24x10= ________

12) 67x100= ________

13) 93x10= ________

14) 47x10= ________

15) 86x1000= ________

16) 67x1000= ________

17) 64x100= ________

18) 20x1000= ________

19) 92x1000= ________

20) 54x10= ________

Name ________________________ Date ________________

Multiplication Multiply by 10, 100 and 1000

Multiply.

1) 80x100= ________

2) 26x1000= ________

3) 67x100= ________

4) 85x1000= ________

5) 88x1000= ________

6) 41x100= ________

7) 37x10= ________

8) 41x1000= ________

9) 61x100= ________

10) 56x1000= ________

11) 63x1000= ________

12) 64x10= ________

13) 42x100= ________

14) 27x100= ________

15) 92x1000= ________

16) 94x1000= ________

17) 22x100= ________

18) 73x1000= ________

19) 65x100= ________

20) 30x1000= ________

Name ______________________ Date ______________

Multiplication Multiply by 10, 100 and 1000

Multiply.

1) $76 \times 1000 =$ ________

2) $82 \times 10 =$ ________

3) $52 \times 10 =$ ________

4) $35 \times 10 =$ ________

5) $40 \times 1000 =$ ________

6) $32 \times 10 =$ ________

7) $19 \times 100 =$ ________

8) $53 \times 10 =$ ________

9) $54 \times 100 =$ ________

10) $75 \times 100 =$ ________

11) $47 \times 100 =$ ________

12) $53 \times 100 =$ ________

13) $31 \times 10 =$ ________

14) $96 \times 1000 =$ ________

15) $64 \times 10 =$ ________

16) $50 \times 100 =$ ________

17) $41 \times 10 =$ ________

18) $65 \times 10 =$ ________

19) $25 \times 100 =$ ________

20) $34 \times 100 =$ ________

SECTION

13

FIND THE MISSING MULTIPLIERS
10, 100, 1000

11 worksheets
20 problems per sheet

Name ______________________________ Date ______________________

Multiplication Find the Missing Multipliers (10, 100 or 1000)

Fill in the blanks with 10, 100 or 1000.

1) 11x ________ =11000

2) 69x ________ =690

3) 97x ________ =97000

4) 90x ________ =900

5) 21x ________ =2100

6) 2x ________ =20

7) 32x ________ =3200

8) 59x ________ =590

9) 3x ________ =30

10) 78x ________ =78000

11) 7x ________ =70

12) 57x ________ =57000

13) 69x ________ =6900

14) 5x ________ =50

15) 74x ________ =740

16) 4x ________ =40

17) 60x ________ =600

18) 79x ________ =7900

19) 42x ________ =420

20) 41x ________ =4100

Name ______________________________ Date ____________________

Multiplication Find the Missing Multipliers (10, 100 or 1000)

Fill in the blanks with 10, 100 or 1000.

1) 73x ________ =73000

2) 8x ________ =80

3) 51x ________ =510

4) 28x ________ =28000

5) 10x ________ =100

6) 17x ________ =1700

7) 23x ________ =23000

8) 48x ________ =480

9) 65x ________ =6500

10) 96x ________ =960

11) 24x ________ =2400

12) 86x ________ =86000

13) 79x ________ =79000

14) 95x ________ =9500

15) 84x ________ =8400

16) 42x ________ =420

17) 15x ________ =15000

18) 98x ________ =98000

19) 43x ________ =430

20) 43x ________ =43000

Name ______________________ Date ______________________

Multiplication Find the Missing Multipliers (10, 100 or 1000)

Fill in the blanks with 10, 100 or 1000.

1) 16x _______ =1600

2) 40x _______ =40000

3) 45x _______ =45000

4) 51x _______ =5100

5) 30x _______ =30000

6) 5x _______ =5000

7) 46x _______ =46000

8) 75x _______ =75000

9) 66x _______ =66000

10) 40x _______ =400

11) 46x _______ =4600

12) 49x _______ =49000

13) 97x _______ =9700

14) 90x _______ =900

15) 74x _______ =74000

16) 45x _______ =4500

17) 50x _______ =500

18) 85x _______ =85000

19) 98x _______ =9800

20) 87x _______ =87000

Name ______________________________ Date ____________________

Multiplication Find the Missing Multipliers (10, 100 or 1000)

Fill in the blanks with 10, 100 or 1000.

1) 67x _______ =67000

2) 72x _______ =7200

3) 26x _______ =260

4) 48x _______ =48000

5) 14x _______ =1400

6) 41x _______ =41000

7) 7x _______ =7000

8) 17x _______ =170

9) 86x _______ =860

10) 52x _______ =5200

11) 97x _______ =9700

12) 28x _______ =28000

13) 23x _______ =230

14) 78x _______ =7800

15) 45x _______ =45000

16) 17x _______ =1700

17) 41x _______ =410

18) 78x _______ =780

19) 81x _______ =810

20) 9x _______ =9000

Name ______________________ Date ______________

Multiplication Find the Missing Multipliers (10, 100 or 1000)

Fill in the blanks with 10, 100 or 1000.

1) 57x _______ =57000

2) 98x _______ =98000

3) 27x _______ =2700

4) 35x _______ =3500

5) 69x _______ =690

6) 31x _______ =3100

7) 62x _______ =620

8) 77x _______ =77000

9) 96x _______ =960

10) 10x _______ =10000

11) 78x _______ =780

12) 67x _______ =6700

13) 93x _______ =930

14) 32x _______ =320

15) 4x _______ =40

16) 76x _______ =76000

17) 63x _______ =63000

18) 79x _______ =79000

19) 9x _______ =900

20) 94x _______ =9400

Name ______________________ Date ______________

Multiplication Find the Missing Multipliers (10, 100 or 1000)

Fill in the blanks with 10, 100 or 1000.

1) 35x ________ =35000

2) 46x ________ =46000

3) 64x ________ =6400

4) 63x ________ =63000

5) 62x ________ =6200

6) 36x ________ =360

7) 5x ________ =500

8) 53x ________ =530

9) 47x ________ =470

10) 16x ________ =1600

11) 40x ________ =40000

12) 85x ________ =850

13) 33x ________ =3300

14) 26x ________ =2600

15) 62x ________ =620

16) 88x ________ =8800

17) 39x ________ =3900

18) 62x ________ =62000

19) 89x ________ =89000

20) 23x ________ =23000

Name ______________________ Date ______________________

Multiplication Find the Missing Multipliers (10, 100 or 1000)

Fill in the blanks with 10, 100 or 1000.

1) 78x _______ =78000

2) 60x _______ =600

3) 80x _______ =80000

4) 14x _______ =1400

5) 76x _______ =7600

6) 46x _______ =460

7) 41x _______ =41000

8) 34x _______ =3400

9) 47x _______ =470

10) 52x _______ =520

11) 58x _______ =580

12) 30x _______ =3000

13) 80x _______ =800

14) 60x _______ =6000

15) 64x _______ =6400

16) 41x _______ =410

17) 84x _______ =840

18) 45x _______ =45000

19) 18x _______ =180

20) 49x _______ =490

Name ________________________ Date ________________

Multiplication Find the Missing Multipliers (10, 100 or 1000)

Fill in the blanks with 10, 100 or 1000.

1) 88x ______ =88000

2) 6x ______ =6000

3) 95x ______ =95000

4) 37x ______ =37000

5) 47x ______ =470

6) 50x ______ =5000

7) 10x ______ =10000

8) 22x ______ =22000

9) 83x ______ =830

10) 43x ______ =43000

11) 56x ______ =56000

12) 54x ______ =540

13) 70x ______ =7000

14) 52x ______ =520

15) 98x ______ =980

16) 16x ______ =1600

17) 67x ______ =670

18) 76x ______ =7600

19) 17x ______ =170

20) 7x ______ =700

Name ______________________ Date ______________

Multiplication Find the Missing Multipliers (10, 100 or 1000)

Fill in the blanks with 10, 100 or 1000.

1) 93x ________ =9300

2) 86x ________ =860

3) 19x ________ =1900

4) 32x ________ =3200

5) 28x ________ =28000

6) 50x ________ =50000

7) 97x ________ =97000

8) 19x ________ =19000

9) 8x ________ =800

10) 11x ________ =110

11) 20x ________ =2000

12) 77x ________ =7700

13) 54x ________ =540

14) 68x ________ =68000

15) 92x ________ =9200

16) 51x ________ =51000

17) 15x ________ =15000

18) 4x ________ =400

19) 88x ________ =8800

20) 95x ________ =950

Name ________________________ Date ____________________

Multiplication Find the Missing Multipliers (10, 100 or 1000)

Fill in the blanks with 10, 100 or 1000.

1) 27x _______ =270

2) 3x _______ =300

3) 33x _______ =330

4) 66x _______ =660

5) 2x _______ =200

6) 80x _______ =800

7) 7x _______ =70

8) 21x _______ =2100

9) 23x _______ =230

10) 58x _______ =580

11) 94x _______ =9400

12) 10x _______ =1000

13) 9x _______ =90

14) 15x _______ =150

15) 98x _______ =98000

16) 86x _______ =86000

17) 47x _______ =47000

18) 4x _______ =400

19) 26x _______ =260

20) 37x _______ =3700

Name ________________________ Date ________________

Multiplication Find the Missing Multipliers (10, 100 or 1000)

Fill in the blanks with 10, 100 or 1000.

1) 61x ______ =610

2) 71x ______ =7100

3) 60x ______ =600

4) 25x ______ =25000

5) 11x ______ =1100

6) 15x ______ =1500

7) 82x ______ =8200

8) 16x ______ =1600

9) 55x ______ =5500

10) 4x ______ =40

11) 25x ______ =2500

12) 89x ______ =8900

13) 39x ______ =3900

14) 42x ______ =42000

15) 22x ______ =2200

16) 48x ______ =4800

17) 46x ______ =460

18) 85x ______ =8500

19) 88x ______ =8800

20) 29x ______ =2900

SECTION

14

MULTIPLICATION
by 10, 100, 1000
with Decimals

11 worksheets
20 problems per sheet

Name ______________________ Date ______________________

Multiplication
Multiply by 10, 100 and 1000 *(with decimals)*

Multiply.

1) .44x10= ______

2) 9.1x100= ______

3) 4.1x100= ______

4) .85x1000= ______

5) .057x10= ______

6) 4.5x1000= ______

7) .022x100= ______

8) .71x100= ______

9) 1.2x100= ______

10) .53x1000= ______

11) 6.3x1000= ______

12) .86x100= ______

13) .058x100= ______

14) .97x100= ______

15) 5.2x1000= ______

16) .12x10= ______

17) .38x100= ______

18) .9x1000= ______

19) 5.3x100= ______

20) .54x10= ______

Name ______________________ Date ______________

Multiplication Multiply by 10, 100 and 1000 (*with decimals*)

Multiply.

1) .9x1000= ______

2) .064x10= ______

3) 3.9x10= ______

4) 4.5x1000= ______

5) .043x10= ______

6) .09x1000= ______

7) 9.1x100= ______

8) 2.3x100= ______

9) 5.2x100= ______

10) .076x10= ______

11) .089x1000= ______

12) 3.9x1000= ______

13) 6.5x100= ______

14) .067x10= ______

15) .039x1000= ______

16) .03x1000= ______

17) 7.7x10= ______

18) 2.9x100= ______

19) .61x1000= ______

20) .18x1000= ______

Name ______________________________ Date ____________________

Multiplication Multiply by 10, 100 and 1000 (*with decimals*)

Multiply.

1) .5x10= ______

2) .5x1000= ______

3) .14x10= ______

4) .057x100= ______

5) .33x1000= ______

6) .094x10= ______

7) .01x100= ______

8) .64x1000= ______

9) .13x100= ______

10) 2.3x100= ______

11) .093x100= ______

12) .47x1000= ______

13) .17x100= ______

14) 6.7x1000= ______

15) .041x1000= ______

16) .68x100= ______

17) .018x100= ______

18) .63x10= ______

19) .26x10= ______

20) .081x100= ______

Name ______________________ Date ______________________

Multiplication Multiply by 10, 100 and 1000 *(with decimals)*

Multiply.

1) .88x10= ________

2) .055x100= ________

3) .032x10= ________

4) 3.1x1000= ________

5) .5x10= ________

6) .34x10= ________

7) 3.6x10= ________

8) .01x1000= ________

9) 1.5x10= ________

10) .021x1000= ________

11) .072x10= ________

12) 2.3x10= ________

13) .063x1000= ________

14) .63x100= ________

15) 6.2x100= ________

16) 2.7x100= ________

17) 1.2x10= ________

18) .73x10= ________

19) 2.3x100= ________

20) .066x10= ________

Name ______________________ Date ______________

Multiplication Multiply by 10, 100 and 1000 *(with decimals)*

Multiply.

1) .014x10= ______

2) .098x100= ______

3) .33x10= ______

4) .038x1000= ______

5) 3.8x100= ______

6) 7.6x1000= ______

7) .045x100= ______

8) .06x10= ______

9) .019x100= ______

10) .064x1000= ______

11) 4.7x1000= ______

12) 6.7x100= ______

13) 1.4x100= ______

14) .079x100= ______

15) 9.7x1000= ______

16) 5.7x1000= ______

17) .032x10= ______

18) .4x10= ______

19) .89x10= ______

20) 6.8x10= ______

Name ______________________________ Date ____________________

Multiplication Multiply by 10, 100 and 1000 *(with decimals)*

Multiply.

1) .74x100= ________

2) 4.7x1000= ________

3) .78x1000= ________

4) 8.9x1000= ________

5) .47x10= ________

6) 8.6x10= ________

7) .45x100= ________

8) 7.6x1000= ________

9) .087x1000= ________

10) .52x10= ________

11) .3x1000= ________

12) .059x10= ________

13) .46x10= ________

14) .027x100= ________

15) .05x1000= ________

16) .73x100= ________

17) .091x10= ________

18) .55x100= ________

19) .98x100= ________

20) .72x100= ________

Name ______________________ Date ______________

Multiplication Multiply by 10, 100 and 1000 (*with decimals*)

Multiply.

1) .042x10= ______

2) .51x1000= ______

3) .22x1000= ______

4) .65x1000= ______

5) .092x10= ______

6) .037x10= ______

7) 7.2x1000= ______

8) .54x100= ______

9) 2.7x10= ______

10) 5.8x1000= ______

11) .52x10= ______

12) .079x10= ______

13) 7.7x1000= ______

14) .79x10= ______

15) 5.3x1000= ______

16) .026x10= ______

17) .09x100= ______

18) 2.1x10= ______

19) .2x1000= ______

20) 1.8x1000= ______

Name ______________________ Date ______________

Multiplication Multiply by 10, 100 and 1000 (*with decimals)*

Multiply.

1) 6.7x10= ______

2) .042x10= ______

3) .73x10= ______

4) .79x100= ______

5) .025x100= ______

6) .04x100= ______

7) .044x100= ______

8) .039x100= ______

9) .61x100= ______

10) .92x10= ______

11) 1.6x10= ______

12) .064x1000= ______

13) .043x100= ______

14) .031x1000= ______

15) .53x10= ______

16) .33x1000= ______

17) 1.4x1000= ______

18) .85x1000= ______

19) .015x1000= ______

20) 5.6x10= ______

Name ______________________ Date ______________________

Multiplication Multiply by 10, 100 and 1000 *(with decimals)*

Multiply.

1) .054x10= ______

2) .2x1000= ______

3) .05x1000= ______

4) .2x10= ______

5) .029x100= ______

6) .46x10= ______

7) 1.8x10= ______

8) .04x10= ______

9) 9.2x1000= ______

10) .43x10= ______

11) 9.4x10= ______

12) 5.9x100= ______

13) .49x100= ______

14) .061x100= ______

15) .16x10= ______

16) 7.5x1000= ______

17) .082x10= ______

18) .7x100= ______

19) .094x1000= ______

20) 8.5x100= ______

Name ______________________ Date ______________

Multiplication Multiply by 10, 100 and 1000 *(with decimals)*

Multiply.

1) .042x100= ______

2) .12x10= ______

3) 6.3x1000= ______

4) .038x10= ______

5) .05x10= ______

6) .008x1000= ______

7) 4.2x100= ______

8) .34x100= ______

9) .32x1000= ______

10) 8.2x100= ______

11) 6.9x100= ______

12) .072x100= ______

13) .36x1000= ______

14) .81x10= ______

15) .45x1000= ______

16) .91x100= ______

17) .051x100= ______

18) .032x10= ______

19) .55x10= ______

20) .076x10= ______

Name ______________________ Date ______________

Multiplication Multiply by 10, 100 and 1000 (*with decimals*)

Multiply.

1) 5.8x10= ______

2) .071x1000= ______

3) .14x1000= ______

4) 6.5x1000= ______

5) .36x10= ______

6) .077x10= ______

7) .019x10= ______

8) 7.6x100= ______

9) .02x1000= ______

10) .41x1000= ______

11) .12x100= ______

12) 5.7x100= ______

13) .024x10= ______

14) .051x1000= ______

15) 4.3x1000= ______

16) 8.4x10= ______

17) .074x10= ______

18) .4x100= ______

19) .77x100= ______

20) .072x1000= ______

SECTION

15

FIND THE MISSING MULTIPLIERS

10, 100, 1000

with Decimals

11 worksheets

20 problems per sheet

Name ______________________ Date ______________

Multiplication Find the Missing Multipliers (10, 100 or 1000)

Fill in the blanks with 10, 100 or 1000.

1) 6.9x _______ =690

2) .19x _______ =190

3) .79x _______ =790

4) .74x _______ =740

5) .097x _______ =.97

6) .02x _______ =20

7) 6.3x _______ =63

8) .42x _______ =4.2

9) .76x _______ =7.6

10) 2.3x _______ =2300

11) 5.4x _______ =540

12) .13x _______ =13

13) .091x _______ =.91

14) .07x _______ =7

15) .058x _______ =5.8

16) 8.4x _______ =840

17) .032x _______ =.32

18) 4.7x _______ =470

19) .77x _______ =77

20) .094x _______ =9.4

Name ______________________ Date ______________

Multiplication Find the Missing Multipliers (10, 100 or 1000)

Fill in the blanks with 10, 100 or 1000.

1) .91x ________ =9.1

2) 8.6x ________ =8600

3) 1.5x ________ =1500

4) 1.6x ________ =1600

5) .92x ________ =92

6) .035x ________ =.35

7) .34x ________ =3.4

8) .023x ________ =.23

9) .088x ________ =8.8

10) .68x ________ =680

11) 1.5x ________ =15

12) .022x ________ =2.2

13) .005x ________ =5

14) .066x ________ =6.6

15) .63x ________ =6.3

16) .021x ________ =21

17) .059x ________ =.59

18) .25x ________ =25

19) .08x ________ =8

20) .62x ________ =62

Name ________________________ Date ________________

Multiplication Find the Missing Multipliers (10, 100 or 1000)

Fill in the blanks with 10, 100 or 1000.

1) .07x ______ =.7

2) 9.3x ______ =9300

3) .67x ______ =6.7

4) .9x ______ =900

5) .062x ______ =6.2

6) .02x ______ =2

7) .04x ______ =4

8) 4.7x ______ =470

9) 8.3x ______ =83

10) .007x ______ =7

11) 6.9x ______ =690

12) .8x ______ =80

13) .36x ______ =3.6

14) 2.6x ______ =260

15) .081x ______ =81

16) .071x ______ =.71

17) 1.4x ______ =14

18) .06x ______ =.6

19) .69x ______ =6.9

20) 4.4x ______ =4400

Name ______________________ Date ______________

Multiplication Find the Missing Multipliers (10, 100 or 1000)

Fill in the blanks with 10, 100 or 1000.

1) .46x _______ =46

2) .51x _______ =51

3) .03x _______ =3

4) .037x _______ =.37

5) .054x _______ =54

6) .071x _______ =.71

7) 8.7x _______ =8700

8) .065x _______ =.65

9) .013x _______ =13

10) 7.9x _______ =79

11) .059x _______ =5.9

12) 8.7x _______ =8700

13) .025x _______ =2.5

14) .079x _______ =.79

15) .064x _______ =64

16) 5.8x _______ =580

17) .9x _______ =900

18) .025x _______ =25

19) .036x _______ =.36

20) .093x _______ =93

Name ______________________ Date ______________

Multiplication Find the Missing Multipliers (10, 100 or 1000)

Fill in the blanks with 10, 100 or 1000.

1) .085x ______ =8.5

2) 9.1x ______ =91

3) .6x ______ =600

4) .55x ______ =550

5) 9.4x ______ =9400

6) .007x ______ =.7

7) .019x ______ =.19

8) .022x ______ =22

9) 8.7x ______ =87

10) .5x ______ =500

11) .019x ______ =1.9

12) 1.8x ______ =180

13) .59x ______ =5.9

14) 1.3x ______ =13

15) 8.5x ______ =8500

16) .06x ______ =60

17) .013x ______ =.13

18) .97x ______ =9.7

19) .28x ______ =280

20) .59x ______ =59

Name ______________________ Date ______________

Multiplication Find the Missing Multipliers (10, 100 or 1000)

Fill in the blanks with 10, 100 or 1000.

1) .98x _______ =9.8

2) .009x _______ =.9

3) 8.5x _______ =85

4) .075x _______ =75

5) .87x _______ =87

6) .35x _______ =3.5

7) 4.7x _______ =47

8) .4x _______ =4

9) 5.2x _______ =5200

10) .43x _______ =430

11) .097x _______ =97

12) 7.1x _______ =710

13) .043x _______ =.43

14) .035x _______ =.35

15) 7.9x _______ =79

16) .048x _______ =48

17) .63x _______ =630

18) .04x _______ =.4

19) .095x _______ =95

20) .087x _______ =87

Name ______________________________ Date ____________________

Multiplication Find the Missing Multipliers (10, 100 or 1000)

Fill in the blanks with 10, 100 or 1000.

1) 4.4x ________ =440

2) 7.1x ________ =710

3) 1.5x ________ =15

4) 5.5x ________ =550

5) .04x ________ =.4

6) .11x ________ =110

7) .002x ________ =.02

8) .97x ________ =9.7

9) .48x ________ =4.8

10) .04x ________ =4

11) .031x ________ =3.1

12) 2.3x ________ =2300

13) .7x ________ =700

14) .19x ________ =19

15) .68x ________ =68

16) .008x ________ =8

17) .29x ________ =29

18) .63x ________ =630

19) .004x ________ =.4

20) .019x ________ =19

Name ________________________________ Date ____________________

Multiplication Find the Missing Multipliers (10, 100 or 1000)

Fill in the blanks with 10, 100 or 1000.

1) .037x _______ =3.7
2) .096x _______ =9.6
3) .07x _______ =.7
4) .87x _______ =870
5) .047x _______ =47
6) .097x _______ =9.7
7) .058x _______ =58
8) .25x _______ =250
9) .053x _______ =5.3
10) 3.8x _______ =380
11) 9.3x _______ =9300
12) .49x _______ =4.9
13) 5.7x _______ =570
14) 8.2x _______ =8200
15) .77x _______ =77
16) .98x _______ =9.8
17) 7.9x _______ =79
18) 5.6x _______ =560
19) .062x _______ =.62
20) .055x _______ =.55

Name ______________________ Date ______________________

Multiplication Find the Missing Multipliers (10, 100 or 1000)

Fill in the blanks with 10, 100 or 1000.

1) .32x _______ =32

2) .068x _______ =68

3) 9.4x _______ =940

4) .91x _______ =910

5) .98x _______ =980

6) .71x _______ =710

7) .48x _______ =480

8) 5.2x _______ =520

9) .062x _______ =62

10) 3.1x _______ =310

11) .04x _______ =.4

12) 8.8x _______ =880

13) .77x _______ =77

14) .97x _______ =970

15) 9.1x _______ =9100

16) 4.6x _______ =4600

17) .086x _______ =8.6

18) .88x _______ =8.8

19) 6.3x _______ =63

20) .098x _______ =.98

Name ______________________________ Date ____________________

Multiplication Find the Missing Multipliers (10, 100 or 1000)

Fill in the blanks with 10, 100 or 1000.

1) .046x ________ =46

2) 2.5x ________ =2500

3) .64x ________ =6.4

4) 7.3x ________ =730

5) .94x ________ =9.4

6) .32x ________ =3.2

7) .067x ________ =6.7

8) .002x ________ =.02

9) .59x ________ =590

10) .034x ________ =34

11) 4.3x ________ =4300

12) .075x ________ =75

13) .084x ________ =84

14) .56x ________ =56

15) 9.5x ________ =9500

16) .57x ________ =5.7

17) .065x ________ =.65

18) 4.4x ________ =4400

19) .036x ________ =.36

20) .032x ________ =3.2

Name ______________________ Date ______________

Multiplication Find the Missing Multipliers (10, 100 or 1000)

Fill in the blanks with 10, 100 or 1000.

1) 9.7x ______ =9700

2) 3.6x ______ =360

3) 6.2x ______ =6200

4) .057x ______ =57

5) .98x ______ =9.8

6) .11x ______ =11

7) 7.3x ______ =73

8) .59x ______ =5.9

9) .04x ______ =40

10) .039x ______ =.39

11) .094x ______ =9.4

12) .085x ______ =85

13) .85x ______ =8.5

14) .25x ______ =25

15) .5x ______ =5

16) .032x ______ =3.2

17) 2.8x ______ =2800

18) .42x ______ =420

19) .32x ______ =320

20) .089x ______ =89

SECTION

16

DIVISION
by 10, 100, 1000

11 worksheets
20 problems per sheet

Name ______________________ Date ______________________

Division Divide by 10, 100 and 1000

Divide.

1) $3200 \div 100 =$ ____

2) $740 \div 10 =$ ____

3) $3800 \div 100 =$ ____

4) $9200 \div 100 =$ ____

5) $62000 \div 1000 =$ ____

6) $180 \div 10 =$ ____

7) $79000 \div 1000 =$ ____

8) $200 \div 10 =$ ____

9) $5000 \div 100 =$ ____

10) $70000 \div 1000 =$ ____

11) $2000 \div 1000 =$ ____

12) $12000 \div 1000 =$ ____

13) $330 \div 10 =$ ____

14) $1500 \div 100 =$ ____

15) $52000 \div 1000 =$ ____

16) $91000 \div 1000 =$ ____

17) $230 \div 10 =$ ____

18) $6500 \div 100 =$ ____

19) $810 \div 10 =$ ____

20) $9300 \div 100 =$ ____

Name ______________________ Date ______________

Division Divide by 10, 100 and 1000

Divide.

1) $78000 \div 1000 =$ ____

2) $2200 \div 100 =$ ____

3) $9300 \div 100 =$ ____

4) $4600 \div 100 =$ ____

5) $7100 \div 100 =$ ____

6) $210 \div 10 =$ ____

7) $85000 \div 1000 =$ ____

8) $190 \div 10 =$ ____

9) $13000 \div 1000 =$ ____

10) $170 \div 10 =$ ____

11) $72000 \div 1000 =$ ____

12) $41000 \div 1000 =$ ____

13) $680 \div 10 =$ ____

14) $79000 \div 1000 =$ ____

15) $45000 \div 1000 =$ ____

16) $8600 \div 100 =$ ____

17) $7700 \div 100 =$ ____

18) $38000 \div 1000 =$ ____

19) $40000 \div 1000 =$ ____

20) $2600 \div 100 =$ ____

Name ______________________ Date ______________

Division Divide by 10, 100 and 1000

Divide.

1) $82000 \div 1000 =$ ____

2) $4300 \div 100 =$ ____

3) $390 \div 10 =$ ____

4) $910 \div 10 =$ ____

5) $74000 \div 1000 =$ ____

6) $1000 \div 100 =$ ____

7) $650 \div 10 =$ ____

8) $91000 \div 1000 =$ ____

9) $46000 \div 1000 =$ ____

10) $1600 \div 100 =$ ____

11) $6800 \div 100 =$ ____

12) $22000 \div 1000 =$ ____

13) $280 \div 10 =$ ____

14) $820 \div 10 =$ ____

15) $3600 \div 100 =$ ____

16) $8100 \div 100 =$ ____

17) $680 \div 10 =$ ____

18) $87000 \div 1000 =$ ____

19) $8800 \div 100 =$ ____

20) $7000 \div 1000 =$ ____

Name ______________________ Date ______________

Division Divide by 10, 100 and 1000

Divide.

1) 950÷10= ____

2) 810÷10= ____

3) 390÷10= ____

4) 64000÷1000= ____

5) 700÷100= ____

6) 590÷10= ____

7) 720÷10= ____

8) 410÷10= ____

9) 14000÷1000= ____

10) 8300÷100= ____

11) 900÷10= ____

12) 90000÷1000= ____

13) 18000÷1000= ____

14) 4200÷100= ____

15) 86000÷1000= ____

16) 24000÷1000= ____

17) 730÷10= ____

18) 1100÷100= ____

19) 540÷10= ____

20) 76000÷1000= ____

Name ______________________ Date ______________

Division Divide by 10, 100 and 1000

Divide.

1) $640 \div 10 =$ ____

2) $17000 \div 1000 =$ ____

3) $5500 \div 100 =$ ____

4) $30000 \div 1000 =$ ____

5) $340 \div 10 =$ ____

6) $18000 \div 1000 =$ ____

7) $7300 \div 100 =$ ____

8) $5300 \div 100 =$ ____

9) $51000 \div 1000 =$ ____

10) $15000 \div 1000 =$ ____

11) $98000 \div 1000 =$ ____

12) $910 \div 10 =$ ____

13) $53000 \div 1000 =$ ____

14) $3000 \div 1000 =$ ____

15) $130 \div 10 =$ ____

16) $5000 \div 1000 =$ ____

17) $500 \div 10 =$ ____

18) $73000 \div 1000 =$ ____

19) $45000 \div 1000 =$ ____

20) $850 \div 10 =$ ____

Name ______________________ Date ______________

Division Divide by 10, 100 and 1000

Divide.

1) 8100÷100= ____

2) 860÷10= ____

3) 1700÷100= ____

4) 1800÷100= ____

5) 8500÷100= ____

6) 7100÷100= ____

7) 680÷10= ____

8) 2600÷100= ____

9) 38000÷1000= ____

10) 5300÷100= ____

11) 76000÷1000= ____

12) 6500÷100= ____

13) 59000÷1000= ____

14) 7900÷100= ____

15) 96000÷1000= ____

16) 37000÷1000= ____

17) 6600÷100= ____

18) 1200÷100= ____

19) 41000÷1000= ____

20) 780÷10= ____

Name ______________________________ Date ________________

Division Divide by 10, 100 and 1000

Divide.

1) 6700÷100= ____

2) 830÷10= ____

3) 7200÷100= ____

4) 30000÷1000= ____

5) 360÷10= ____

6) 530÷10= ____

7) 8400÷100= ____

8) 34000÷1000= ____

9) 370÷10= ____

10) 840÷10= ____

11) 160÷10= ____

12) 95000÷1000= ____

13) 5000÷1000= ____

14) 1100÷100= ____

15) 8500÷100= ____

16) 31000÷1000= ____

17) 3100÷100= ____

18) 28000÷1000= ____

19) 3300÷100= ____

20) 38000÷1000= ____

Name ____________________ Date ____________________

Division Divide by 10, 100 and 1000

Divide.

1) $1900 \div 100 =$ ____

2) $500 \div 10 =$ ____

3) $5100 \div 100 =$ ____

4) $670 \div 10 =$ ____

5) $9400 \div 100 =$ ____

6) $820 \div 10 =$ ____

7) $280 \div 10 =$ ____

8) $7300 \div 100 =$ ____

9) $21000 \div 1000 =$ ____

10) $140 \div 10 =$ ____

11) $64000 \div 1000 =$ ____

12) $6500 \div 100 =$ ____

13) $650 \div 10 =$ ____

14) $840 \div 10 =$ ____

15) $95000 \div 1000 =$ ____

16) $230 \div 10 =$ ____

17) $340 \div 10 =$ ____

18) $52000 \div 1000 =$ ____

19) $50000 \div 1000 =$ ____

20) $2400 \div 100 =$ ____

Name ______________________ Date ______________

Division Divide by 10, 100 and 1000

Divide.

1) $7600 \div 100 =$ ____

2) $270 \div 10 =$ ____

3) $5000 \div 100 =$ ____

4) $35000 \div 1000 =$ ____

5) $400 \div 100 =$ ____

6) $250 \div 10 =$ ____

7) $23000 \div 1000 =$ ____

8) $6800 \div 100 =$ ____

9) $8400 \div 100 =$ ____

10) $840 \div 10 =$ ____

11) $90000 \div 1000 =$ ____

12) $40000 \div 1000 =$ ____

13) $210 \div 10 =$ ____

14) $7700 \div 100 =$ ____

15) $11000 \div 1000 =$ ____

16) $6400 \div 100 =$ ____

17) $3700 \div 100 =$ ____

18) $740 \div 10 =$ ____

19) $4400 \div 100 =$ ____

20) $6500 \div 100 =$ ____

Name ______________________ Date ______________

Division Divide by 10, 100 and 1000

Divide.

1) 82000÷1000= ____

2) 1600÷100= ____

3) 62000÷1000= ____

4) 23000÷1000= ____

5) 420÷10= ____

6) 27000÷1000= ____

7) 7100÷100= ____

8) 950÷10= ____

9) 6400÷100= ____

10) 2400÷100= ____

11) 52000÷1000= ____

12) 340÷10= ____

13) 3400÷100= ____

14) 270÷10= ____

15) 1700÷100= ____

16) 930÷10= ____

17) 700÷100= ____

18) 85000÷1000= ____

19) 4100÷100= ____

20) 8900÷100= ____

Name ______________________ Date ______________

Division Divide by 10, 100 and 1000

Divide.

1) 610÷10= ____

2) 3700÷100= ____

3) 2900÷100= ____

4) 32000÷1000= ____

5) 930÷10= ____

6) 14000÷1000= ____

7) 35000÷1000= ____

8) 560÷10= ____

9) 97000÷1000= ____

10) 470÷10= ____

11) 1500÷100= ____

12) 88000÷1000= ____

13) 82000÷1000= ____

14) 690÷10= ____

15) 70000÷1000= ____

16) 57000÷1000= ____

17) 510÷10= ____

18) 250÷10= ____

19) 36000÷1000= ____

20) 380÷10= ____

SECTION

17

FIND THE MISSING DIVISORS
10, 100, 1000

11 worksheets
20 problems per sheet

Name ______________________ Date ______________

Division Find the Missing Divisors (10, 100 or 1000)

Fill in the blanks with 10, 100 or 1000.

1) 38000÷ ________ =38

2) 490÷ ________ =49

3) 3300÷ ________ =33

4) 97000÷ ________ =97

5) 600÷ ________ =6

6) 6200÷ ________ =62

7) 300÷ ________ =30

8) 17000÷ ________ =17

9) 8100÷ ________ =81

10) 3600÷ ________ =36

11) 3800÷ ________ =38

12) 740÷ ________ =74

13) 21000÷ ________ =21

14) 58000÷ ________ =58

15) 2500÷ ________ =25

16) 200÷ ________ =2

17) 240÷ ________ =24

18) 760÷ ________ =76

19) 93000÷ ________ =93

20) 6100÷ ________ =61

Name ______________________ Date ______________________

Division Find the Missing Divisors (10, 100 or 1000)

Fill in the blanks with 10, 100 or 1000.

1) 6300÷ ________ =63

2) 4500÷ ________ =45

3) 750÷ ________ =75

4) 63000÷ ________ =63

5) 940÷ ________ =94

6) 72000÷ ________ =72

7) 670÷ ________ =67

8) 320÷ ________ =32

9) 2200÷ ________ =22

10) 2900÷ ________ =29

11) 680÷ ________ =68

12) 95000÷ ________ =95

13) 540÷ ________ =54

14) 250÷ ________ =25

15) 8400÷ ________ =84

16) 83000÷ ________ =83

17) 67000÷ ________ =67

18) 240÷ ________ =24

19) 40000÷ ________ =40

20) 85000÷ ________ =85

Name ______________________ Date ______________

Division Find the Missing Divisors (10, 100 or 1000)

Fill in the blanks with 10, 100 or 1000.

1) 17000÷ _______ =17

2) 4200÷ _______ =42

3) 5900÷ _______ =59

4) 90000÷ _______ =90

5) 1000÷ _______ =10

6) 340÷ _______ =34

7) 39000÷ _______ =39

8) 450÷ _______ =45

9) 24000÷ _______ =24

10) 9600÷ _______ =96

11) 580÷ _______ =58

12) 430÷ _______ =43

13) 86000÷ _______ =86

14) 8600÷ _______ =86

15) 8800÷ _______ =88

16) 91000÷ _______ =91

17) 42000÷ _______ =42

18) 3500÷ _______ =35

19) 19000÷ _______ =19

20) 88000÷ _______ =88

Name ______________________ Date ______________________

Division Find the Missing Divisors (10, 100 or 1000)

Fill in the blanks with 10, 100 or 1000.

1) 390÷ _______ =39

2) 69000÷ _______ =69

3) 6000÷ _______ =60

4) 340÷ _______ =34

5) 6100÷ _______ =61

6) 6600÷ _______ =66

7) 2000÷ _______ =20

8) 42000÷ _______ =42

9) 87000÷ _______ =87

10) 60000÷ _______ =60

11) 920÷ _______ =92

12) 30000÷ _______ =30

13) 440÷ _______ =44

14) 3300÷ _______ =33

15) 43000÷ _______ =43

16) 71000÷ _______ =71

17) 650÷ _______ =65

18) 230÷ _______ =23

19) 3500÷ _______ =35

20) 980÷ _______ =98

Name ______________________ Date ______________

Division Find the Missing Divisors (10, 100 or 1000)

Fill in the blanks with 10, 100 or 1000.

1) 5600÷ ________ =56

2) 26000÷ ________ =26

3) 44000÷ ________ =44

4) 580÷ ________ =58

5) 6600÷ ________ =660

6) 8200÷ ________ =82

7) 7200÷ ________ =720

8) 250÷ ________ =25

9) 5000÷ ________ =50

10) 900÷ ________ =9

11) 3100÷ ________ =310

12) 46000÷ ________ =46

13) 900÷ ________ =90

14) 860÷ ________ =86

15) 170÷ ________ =17

16) 54000÷ ________ =54

17) 740÷ ________ =74

18) 8000÷ ________ =800

19) 700÷ ________ =70

20) 400÷ ________ =40

Name ______________________________ Date ______________________

Division Find the Missing Divisors (10, 100 or 1000)

Fill in the blanks with 10, 100 or 1000.

1) 8300÷ ________ =83

2) 290÷ ________ =29

3) 87000÷ ________ =87

4) 6000÷ ________ =60

5) 910÷ ________ =91

6) 7900÷ ________ =79

7) 82000÷ ________ =82

8) 960÷ ________ =96

9) 77000÷ ________ =77

10) 49000÷ ________ =49

11) 35000÷ ________ =35

12) 28000÷ ________ =28

13) 41000÷ ________ =41

14) 460÷ ________ =46

15) 65000÷ ________ =65

16) 70000÷ ________ =70

17) 74000÷ ________ =74

18) 6800÷ ________ =68

19) 36000÷ ________ =36

20) 9600÷ ________ =96

Name ________________________________ Date ____________________

Division Find the Missing Divisors (10, 100 or 1000)

Fill in the blanks with 10, 100 or 1000.

1) 2700÷ ________ =27

2) 44000÷ ________ =44

3) 2500÷ ________ =25

4) 2000÷ ________ =2

5) 97000÷ ________ =97

6) 150÷ ________ =15

7) 95000÷ ________ =95

8) 410÷ ________ =41

9) 920÷ ________ =92

10) 6100÷ ________ =61

11) 9800÷ ________ =98

12) 23000÷ ________ =23

13) 820÷ ________ =82

14) 9700÷ ________ =97

15) 89000÷ ________ =89

16) 310÷ ________ =31

17) 41000÷ ________ =41

18) 9200÷ ________ =92

19) 53000÷ ________ =53

20) 6500÷ ________ =65

Name ______________________ Date ______________

Division Find the Missing Divisors (10, 100 or 1000)

Fill in the blanks with 10, 100 or 1000.

1) 20000÷ ________ =20

2) 68000÷ ________ =68

3) 1500÷ ________ =15

4) 350÷ ________ =35

5) 7500÷ ________ =75

6) 36000÷ ________ =36

7) 61000÷ ________ =61

8) 490÷ ________ =49

9) 4600÷ ________ =46

10) 90000÷ ________ =90

11) 31000÷ ________ =31

12) 400÷ ________ =4

13) 5700÷ ________ =57

14) 6600÷ ________ =66

15) 5200÷ ________ =52

16) 8500÷ ________ =85

17) 890÷ ________ =89

18) 3700÷ ________ =37

19) 46000÷ ________ =46

20) 9400÷ ________ =94

Name ______________________ Date ______________________

Division Find the Missing Divisors (10, 100 or 1000)

Fill in the blanks with 10, 100 or 1000.

1) 7100÷ ________ =71

2) 2000÷ ________ =20

3) 7800÷ ________ =78

4) 96000÷ ________ =96

5) 330÷ ________ =33

6) 290÷ ________ =29

7) 530÷ ________ =53

8) 2200÷ ________ =22

9) 95000÷ ________ =95

10) 140÷ ________ =14

11) 4000÷ ________ =40

12) 40000÷ ________ =40

13) 3300÷ ________ =33

14) 720÷ ________ =72

15) 510÷ ________ =51

16) 13000÷ ________ =13

17) 910÷ ________ =91

18) 4700÷ ________ =47

19) 260÷ ________ =26

20) 19000÷ ________ =19

Name ________________________ Date ________________

Division Find the Missing Divisors (10, 100 or 1000)

Fill in the blanks with 10, 100 or 1000.

1) 920÷ ______ =92

2) 8300÷ ______ =83

3) 8700÷ ______ =87

4) 500÷ ______ =50

5) 69000÷ ______ =69

6) 90000÷ ______ =90

7) 130÷ ______ =13

8) 800÷ ______ =80

9) 5300÷ ______ =53

10) 9500÷ ______ =95

11) 940÷ ______ =94

12) 89000÷ ______ =89

13) 7600÷ ______ =76

14) 760÷ ______ =76

15) 630÷ ______ =63

16) 5700÷ ______ =57

17) 3100÷ ______ =31

18) 9100÷ ______ =91

19) 93000÷ ______ =93

20) 83000÷ ______ =83

Name ______________________ Date ______________________

Division Find the Missing Divisors (10, 100 or 1000)

Fill in the blanks with 10, 100 or 1000.

1) 9400÷ ________ =94

2) 16000÷ ________ =16

3) 490÷ ________ =49

4) 95000÷ ________ =95

5) 500÷ ________ =5

6) 650÷ ________ =65

7) 3200÷ ________ =32

8) 49000÷ ________ =49

9) 63000÷ ________ =63

10) 940÷ ________ =94

11) 8200÷ ________ =82

12) 5100÷ ________ =51

13) 400÷ ________ =40

14) 210÷ ________ =21

15) 15000÷ ________ =15

16) 9300÷ ________ =93

17) 810÷ ________ =81

18) 1800÷ ________ =18

19) 5800÷ ________ =58

20) 4300÷ ________ =43

SECTION

18

DIVISION
by 10, 100, 1000
with Decimals

11 worksheets
20 problems per sheet

Name ______________________ Date ______________

Division Divide by 10, 100 and 1000 *(with decimals)*

Divide.

1) 9.2÷100= ______

2) .75÷10= ______

3) 3.5÷10= ______

4) 64÷1000= ______

5) 8.6÷100= ______

6) 67÷100= ______

7) 44÷1000= ______

8) 750÷100= ______

9) 6.8÷10= ______

10) .7÷100= ______

11) 9.4÷100= ______

12) 2.1÷100= ______

13) 33÷1000= ______

14) .36÷10= ______

15) 9.5÷100= ______

16) 6100÷1000= ______

17) 8.9÷100= ______

18) 430÷100= ______

19) 280÷100= ______

20) 760÷1000= ______

Name ____________________ Date ____________________

Division Divide by 10, 100 and 1000 (*with decimals*)

Divide.

1) $20 \div 100 =$ ______

2) $94 \div 100 =$ ______

3) $8.6 \div 100 =$ ______

4) $96 \div 1000 =$ ______

5) $61 \div 100 =$ ______

6) $470 \div 100 =$ ______

7) $.89 \div 10 =$ ______

8) $260 \div 100 =$ ______

9) $120 \div 1000 =$ ______

10) $5 \div 100 =$ ______

11) $7.3 \div 100 =$ ______

12) $13 \div 1000 =$ ______

13) $42 \div 10 =$ ______

14) $6.8 \div 10 =$ ______

15) $65 \div 1000 =$ ______

16) $7800 \div 1000 =$ ______

17) $680 \div 1000 =$ ______

18) $28 \div 1000 =$ ______

19) $70 \div 100 =$ ______

20) $.42 \div 10 =$ ______

Name ______________________ Date ______________

Division Divide by 10, 100 and 1000 (*with decimals)*

Divide.

1) $18 \div 100 =$ ______

2) $8.4 \div 10 =$ ______

3) $48 \div 1000 =$ ______

4) $460 \div 1000 =$ ______

5) $9.2 \div 10 =$ ______

6) $98 \div 100 =$ ______

7) $94 \div 100 =$ ______

8) $.74 \div 10 =$ ______

9) $53 \div 1000 =$ ______

10) $90 \div 100 =$ ______

11) $7900 \div 1000 =$ ______

12) $20 \div 100 =$ ______

13) $4300 \div 1000 =$ ______

14) $.32 \div 10 =$ ______

15) $980 \div 1000 =$ ______

16) $150 \div 1000 =$ ______

17) $.03 \div 10 =$ ______

18) $44 \div 1000 =$ ______

19) $930 \div 100 =$ ______

20) $.79 \div 10 =$ ______

Name ______________________ Date ______________

Division Divide by 10, 100 and 1000 (*with decimals*)

Divide.

1) $.7 \div 100 =$ ______

2) $930 \div 100 =$ ______

3) $940 \div 100 =$ ______

4) $2.7 \div 10 =$ ______

5) $43 \div 10 =$ ______

6) $68 \div 1000 =$ ______

7) $2.7 \div 100 =$ ______

8) $53 \div 100 =$ ______

9) $.65 \div 10 =$ ______

10) $63 \div 10 =$ ______

11) $18 \div 10 =$ ______

12) $64 \div 100 =$ ______

13) $590 \div 100 =$ ______

14) $80 \div 1000 =$ ______

15) $17 \div 100 =$ ______

16) $5.9 \div 10 =$ ______

17) $.42 \div 10 =$ ______

18) $.6 \div 10 =$ ______

19) $90 \div 100 =$ ______

20) $1.3 \div 10 =$ ______

Name ______________________________ Date ____________________

Division Divide by 10, 100 and 1000 (*with decimals*)

Divide.

1) 640÷1000= ______

2) 4.5÷100= ______

3) .67÷10= ______

4) .03÷10= ______

5) 52÷1000= ______

6) 83÷100= ______

7) .47÷10= ______

8) 960÷1000= ______

9) 9300÷1000= ______

10) 2900÷1000= ______

11) 600÷1000= ______

12) 950÷1000= ______

13) 8.8÷100= ______

14) 78÷100= ______

15) 8.5÷10= ______

16) 7.5÷100= ______

17) .85÷10= ______

18) 5200÷1000= ______

19) 24÷1000= ______

20) .8÷100= ______

Name ______________________ Date ______________

Division Divide by 10, 100 and 1000 (*with decimals)*

Divide.

1) 22÷1000= ______

2) 6.5÷10= ______

3) 95÷100= ______

4) 2100÷1000= ______

5) .89÷10= ______

6) 340÷100= ______

7) 70÷100= ______

8) 470÷100= ______

9) 52÷1000= ______

10) 500÷1000= ______

11) 370÷1000= ______

12) 670÷1000= ______

13) 37÷10= ______

14) 4.4÷10= ______

15) .76÷10= ______

16) 32÷1000= ______

17) 3.8÷100= ______

18) 83÷1000= ______

19) 4.6÷100= ______

20) 6÷1000= ______

Name ______________________ Date ______________

Division Divide by 10, 100 and 1000 *(with decimals)*

Divide.

1) $64 \div 10 =$ ______

2) $25 \div 1000 =$ ______

3) $90 \div 1000 =$ ______

4) $.07 \div 10 =$ ______

5) $1.2 \div 10 =$ ______

6) $4200 \div 1000 =$ ______

7) $30 \div 100 =$ ______

8) $71 \div 10 =$ ______

9) $710 \div 100 =$ ______

10) $4400 \div 1000 =$ ______

11) $.47 \div 10 =$ ______

12) $74 \div 1000 =$ ______

13) $2.2 \div 10 =$ ______

14) $30 \div 100 =$ ______

15) $76 \div 100 =$ ______

16) $570 \div 100 =$ ______

17) $320 \div 1000 =$ ______

18) $.49 \div 10 =$ ______

19) $5900 \div 1000 =$ ______

20) $7500 \div 1000 =$ ______

Name ______________________ Date ______________________

Division Divide by 10, 100 and 1000 (*with decimals*)

Divide.

1) $290 \div 100 =$ ______

2) $.22 \div 10 =$ ______

3) $6.3 \div 10 =$ ______

4) $56 \div 1000 =$ ______

5) $8.7 \div 10 =$ ______

6) $540 \div 1000 =$ ______

7) $82 \div 10 =$ ______

8) $93 \div 10 =$ ______

9) $.59 \div 10 =$ ______

10) $4 \div 100 =$ ______

11) $86 \div 100 =$ ______

12) $4.7 \div 10 =$ ______

13) $630 \div 1000 =$ ______

14) $8700 \div 1000 =$ ______

15) $.48 \div 10 =$ ______

16) $65 \div 100 =$ ______

17) $830 \div 100 =$ ______

18) $750 \div 1000 =$ ______

19) $9400 \div 1000 =$ ______

20) $.43 \div 10 =$ ______

Name ______________________________ Date ____________________

Division Divide by 10, 100 and 1000 (*with decimals)*

Divide.

1) $.17 \div 10 =$ ______

2) $18 \div 100 =$ ______

3) $44 \div 100 =$ ______

4) $92 \div 10 =$ ______

5) $840 \div 1000 =$ ______

6) $83 \div 10 =$ ______

7) $2.9 \div 100 =$ ______

8) $5400 \div 1000 =$ ______

9) $170 \div 1000 =$ ______

10) $1.9 \div 10 =$ ______

11) $.94 \div 10 =$ ______

12) $16 \div 100 =$ ______

13) $70 \div 100 =$ ______

14) $67 \div 10 =$ ______

15) $7.1 \div 10 =$ ______

16) $32 \div 100 =$ ______

17) $500 \div 1000 =$ ______

18) $82 \div 10 =$ ______

19) $2 \div 100 =$ ______

20) $.5 \div 10 =$ ______

Name ____________________ Date ____________________

Division Divide by 10, 100 and 1000 *(with decimals)*

Divide.

1) 37÷10= ______

2) 44÷1000= ______

3) 34÷10= ______

4) 46÷100= ______

5) 93÷100= ______

6) 8.9÷10= ______

7) 62÷100= ______

8) 9÷100= ______

9) 27÷1000= ______

10) 1.7÷10= ______

11) 220÷100= ______

12) 13÷10= ______

13) 6200÷1000= ______

14) 9.5÷100= ______

15) 470÷100= ______

16) 81÷100= ______

17) 2800÷1000= ______

18) 940÷1000= ______

19) .08÷10= ______

20) 1700÷1000= ______

Name ______________________ Date ______________

Division Divide by 10, 100 and 1000 (*with decimals)*

Divide.

1) $2900 \div 1000 =$ ______

2) $.79 \div 10 =$ ______

3) $58 \div 10 =$ ______

4) $.48 \div 10 =$ ______

5) $.56 \div 10 =$ ______

6) $37 \div 10 =$ ______

7) $1400 \div 1000 =$ ______

8) $110 \div 100 =$ ______

9) $70 \div 100 =$ ______

10) $8.4 \div 10 =$ ______

11) $38 \div 100 =$ ______

12) $710 \div 1000 =$ ______

13) $250 \div 1000 =$ ______

14) $400 \div 1000 =$ ______

15) $2 \div 10 =$ ______

16) $50 \div 1000 =$ ______

17) $.27 \div 10 =$ ______

18) $78 \div 10 =$ ______

19) $96 \div 1000 =$ ______

20) $.55 \div 10 =$ ______

SECTION

19

FIND THE MISSING DIVISORS

10, 100, 1000 with Decimals

11 worksheets
20 problems per sheet

Name ______________________________ Date ____________________

Division Find the Missing Divisors (10, 100 or 1000)

Fill in the blanks with 10, 100 or 1000.

1) 570÷ ________ =.57

2) 8.5÷ ________ =.085

3) 6800÷ ________ =6.8

4) 82÷ ________ =.082

5) 6800÷ ________ =6.8

6) 5.1÷ ________ =.051

7) 620÷ ________ =.62

8) 19÷ ________ =1.9

9) 3÷ ________ =.003

10) 5400÷ ________ =5.4

11) 42÷ ________ =4.2

12) 7600÷ ________ =7.6

13) 63÷ ________ =6.3

14) 9400÷ ________ =9.4

15) 1.4÷ ________ =.14

16) 290÷ ________ =2.9

17) 53÷ ________ =5.3

18) 40÷ ________ =.04

19) 6.8÷ ________ =.68

20) 90÷ ________ =.9

Name ______________________________ Date ____________________

Division Find the Missing Divisors (10, 100 or 1000)

Fill in the blanks with 10, 100 or 1000.

1) 24÷ ________ =2.4

2) 77÷ ________ =.77

3) .34÷ ________ =.034

4) 780÷ ________ =7.8

5) 97÷ ________ =.097

6) 980÷ ________ =9.8

7) 340÷ ________ =.34

8) 2.9÷ ________ =.029

9) 6÷ ________ =.06

10) 90÷ ________ =.9

11) 2.4÷ ________ =.024

12) 8.1÷ ________ =.81

13) 63÷ ________ =.63

14) 77÷ ________ =.077

15) 66÷ ________ =6.6

16) .33÷ ________ =.033

17) 6÷ ________ =.6

18) 48÷ ________ =.48

19) 250÷ ________ =.25

20) 78÷ ________ =.78

Name ________________________________ Date ____________________

Division Find the Missing Divisors (10, 100 or 1000)

Fill in the blanks with 10, 100 or 1000.

1) 2.8÷ ______ =.028

2) 27÷ ______ =.027

3) 66÷ ______ =.066

4) 280÷ ______ =2.8

5) 380÷ ______ =.38

6) 980÷ ______ =9.8

7) .09÷ ______ =.009

8) 830÷ ______ =8.3

9) 17÷ ______ =1.7

10) .12÷ ______ =.012

11) 84÷ ______ =.84

12) 28÷ ______ =2.8

13) 2200÷ ______ =2.2

14) 4.6÷ ______ =.46

15) 61÷ ______ =6.1

16) 400÷ ______ =.4

17) 83÷ ______ =.83

18) 9.1÷ ______ =.91

19) .71÷ ______ =.071

20) 55÷ ______ =5.5

Name ______________________ Date ______________________

Division Find the Missing Divisors (10, 100 or 1000)

Fill in the blanks with 10, 100 or 1000.

1) 370÷ _______ =3.7

2) 4.5÷ _______ =.45

3) 720÷ _______ =7.2

4) 65÷ _______ =.065

5) 5÷ _______ =.5

6) 2.6÷ _______ =.026

7) 5600÷ _______ =5.6

8) 29÷ _______ =.29

9) 1100÷ _______ =1.1

10) 28÷ _______ =.28

11) 30÷ _______ =.03

12) 150÷ _______ =1.5

13) 6.7÷ _______ =.67

14) 34÷ _______ =3.4

15) 80÷ _______ =.08

16) .3÷ _______ =.03

17) 1400÷ _______ =1.4

18) 6600÷ _______ =6.6

19) 25÷ _______ =.25

20) 62÷ _______ =.62

Name ______________________ Date ______________

Division Find the Missing Divisors (10, 100 or 1000)

Fill in the blanks with 10, 100 or 1000.

1) 93÷ ______ =.93

2) 2÷ ______ =.02

3) .84÷ ______ =.084

4) 5.5÷ ______ =.055

5) 90÷ ______ =.9

6) 7.3÷ ______ =.073

7) 4.9÷ ______ =.49

8) 180÷ ______ =1.8

9) 29÷ ______ =2.9

10) 3400÷ ______ =3.4

11) 11÷ ______ =.011

12) 64÷ ______ =.64

13) 67÷ ______ =.067

14) 360÷ ______ =3.6

15) 460÷ ______ =4.6

16) 52÷ ______ =.052

17) .41÷ ______ =.041

18) 58÷ ______ =.058

19) 7.1÷ ______ =.071

20) 1.8÷ ______ =.018

Name ______________________ Date ______________

Division Find the Missing Divisors (10, 100 or 1000)

Fill in the blanks with 10, 100 or 1000.

1) 88÷ _______ =.88

2) 8700÷ _______ =8.7

3) 810÷ _______ =8.1

4) 80÷ _______ =.08

5) 530÷ _______ =.53

6) 6.8÷ _______ =.68

7) 940÷ _______ =.94

8) 4.6÷ _______ =.046

9) .31÷ _______ =.031

10) 79÷ _______ =7.9

11) 3200÷ _______ =3.2

12) .7÷ _______ =.07

13) 3.7÷ _______ =.037

14) .13÷ _______ =.013

15) .73÷ _______ =.073

16) 3.2÷ _______ =.032

17) 120÷ _______ =1.2

18) 45÷ _______ =.045

19) 98÷ _______ =.098

20) .04÷ _______ =.004

Name ______________________ Date ______________

Division Find the Missing Divisors (10, 100 or 1000)

Fill in the blanks with 10, 100 or 1000.

1) 440÷ _______ =4.4

2) 320÷ _______ =3.2

3) 600÷ _______ =.6

4) 930÷ _______ =9.3

5) 920÷ _______ =9.2

6) 390÷ _______ =.39

7) 59÷ _______ =.59

8) 5.4÷ _______ =.054

9) 9.8÷ _______ =.98

10) 48÷ _______ =.048

11) 6.5÷ _______ =.065

12) 900÷ _______ =.9

13) 93÷ _______ =9.3

14) 5800÷ _______ =5.8

15) 2÷ _______ =.02

16) 9.5÷ _______ =.95

17) 10÷ _______ =.01

18) 2300÷ _______ =2.3

19) 820÷ _______ =8.2

20) .02÷ _______ =.002

Name ______________________ Date ______________________

Division Find the Missing Divisors (10, 100 or 1000)

Fill in the blanks with 10, 100 or 1000.

1) 2÷ _______ =.02

2) 18÷ _______ =.18

3) 8800÷ _______ =8.8

4) 6900÷ _______ =6.9

5) 93÷ _______ =.093

6) 110÷ _______ =.11

7) 930÷ _______ =9.3

8) 13÷ _______ =.013

9) 31÷ _______ =.031

10) 44÷ _______ =.044

11) 1.5÷ _______ =.015

12) .57÷ _______ =.057

13) .87÷ _______ =.087

14) 11÷ _______ =.011

15) 96÷ _______ =.096

16) 240÷ _______ =.24

17) 870÷ _______ =.87

18) 170÷ _______ =.17

19) 1.1÷ _______ =.011

20) 3900÷ _______ =3.9

Name ______________________ Date ______________________

Division Find the Missing Divisors (10, 100 or 1000)

Fill in the blanks with 10, 100 or 1000.

1) 8.4÷ _______ =.84

2) 3.3÷ _______ =.033

3) 37÷ _______ =.037

4) 45÷ _______ =.45

5) .27÷ _______ =.027

6) 95÷ _______ =9.5

7) 43÷ _______ =4.3

8) 5.2÷ _______ =.052

9) 290÷ _______ =2.9

10) 28÷ _______ =.28

11) 29÷ _______ =.029

12) 8800÷ _______ =8.8

13) 450÷ _______ =.45

14) 480÷ _______ =.48

15) 1500÷ _______ =1.5

16) 4.1÷ _______ =.041

17) .38÷ _______ =.038

18) 4.7÷ _______ =.047

19) 7÷ _______ =.07

20) 9.3÷ _______ =.093

Name ______________________ Date ______________

Division Find the Missing Divisors (10, 100 or 1000)

Fill in the blanks with 10, 100 or 1000.

1) 5.7÷ ________ =.057

2) 790÷ ________ =7.9

3) 60÷ ________ =.06

4) 34÷ ________ =3.4

5) 560÷ ________ =5.6

6) 330÷ ________ =.33

7) 3.8÷ ________ =.038

8) .43÷ ________ =.043

9) 4300÷ ________ =4.3

10) 78÷ ________ =.78

11) 1800÷ ________ =1.8

12) 1.9÷ ________ =.019

13) 9400÷ ________ =9.4

14) 42÷ ________ =4.2

15) .9÷ ________ =.09

16) .31÷ ________ =.031

17) 5.6÷ ________ =.56

18) 6.4÷ ________ =.064

19) 930÷ ________ =.93

20) 70÷ ________ =.7

Name ______________________ Date ______________

Division Find the Missing Divisors (10, 100 or 1000)

Fill in the blanks with 10, 100 or 1000.

1) 3.7÷ ______ =.037

2) .8÷ ______ =.08

3) 8400÷ ______ =8.4

4) 2500÷ ______ =2.5

5) 43÷ ______ =4.3

6) 9.4÷ ______ =.094

7) 1800÷ ______ =1.8

8) 5300÷ ______ =5.3

9) 84÷ ______ =.084

10) 5.1÷ ______ =.051

11) 1.7÷ ______ =.017

12) 36÷ ______ =.36

13) 56÷ ______ =.056

14) .59÷ ______ =.059

15) 74÷ ______ =7.4

16) .7÷ ______ =.07

17) 280÷ ______ =.28

18) .13÷ ______ =.013

19) 55÷ ______ =.055

20) 75÷ ______ =.075

SOLUTIONS

S

SOLUTIONS TO PROBLEMS

Sections 1 - 19

PAGE: 2

Multiplication Facts

Multiply.

1) 6 x2 12	2) 2 x9 18	3) 3 x7 21	4) 8 x2 16	5) 2 x1 2
6) 8 x9 72	7) 12 x7 84	8) 3 x3 9	9) 9 x4 36	10) 8 x5 40
11) 6 x0 0	12) 7 x5 35	13) 10 x5 50	14) 9 x2 18	15) 10 x6 60
16) 11 x9 99	17) 5 x8 40	18) 10 x11 110	19) 12 x8 96	20) 10 x9 90

PAGE: 3

Multiplication Facts

Multiply.

1) 8 x7 56	2) 2 x9 18	3) 12 x11 132	4) 11 x10 110	5) 8 x9 72
6) 7 x9 63	7) 2 x5 10	8) 6 x7 42	9) 7 x0 0	10) 6 x4 24
11) 6 x1 6	12) 10 x3 30	13) 2 x8 16	14) 5 x3 15	15) 11 x2 22
16) 3 x3 9	17) 12 x7 84	18) 7 x7 49	19) 7 x2 14	20) 9 x9 81

PAGE: 4

Multiplication Facts

Multiply.

1) 11 x3 33	2) 10 x12 120	3) 12 x5 60	4) 8 x5 40	5) 8 x9 72
6) 12 x6 72	7) 3 x5 15	8) 4 x1 4	9) 11 x4 44	10) 4 x4 16
11) 5 x9 45	12) 5 x5 25	13) 5 x0 0	14) 8 x7 56	15) 8 x8 64
16) 10 x3 30	17) 10 x8 80	18) 2 x9 18	19) 6 x2 12	20) 4 x5 20

PAGE: 5

Multiplication Facts

Multiply.

1) 7 x6 42	2) 5 x1 5	3) 2 x7 14	4) 7 x5 35	5) 9 x7 63
6) 4 x8 32	7) 7 x8 56	8) 2 x6 12	9) 10 x12 120	10) 11 x3 33
11) 11 x5 55	12) 6 x4 24	13) 9 x9 81	14) 8 x0 0	15) 8 x4 32
16) 6 x7 42	17) 5 x8 40	18) 3 x3 9	19) 4 x5 20	20) 5 x2 10

PAGE: 6

Multiplication Facts

Multiply.

1) 2 x6 12	2) 3 x7 21	3) 7 x1 7	4) 10 x11 110	5) 6 x0 0
6) 6 x4 24	7) 6 x3 18	8) 10 x2 20	9) 12 x6 72	10) 11 x4 44
11) 11 x5 55	12) 5 x6 30	13) 7 x6 42	14) 4 x2 8	15) 3 x8 24
16) 9 x3 27	17) 7 x2 14	18) 12 x7 84	19) 6 x2 12	20) 4 x4 16

PAGE: 7

Multiplication Facts

Multiply.

1) 5 x8 40	2) 2 x1 2	3) 4 x5 20	4) 6 x0 0	5) 7 x6 42
6) 3 x3 9	7) 10 x2 20	8) 2 x6 12	9) 8 x4 32	10) 2 x5 10
11) 7 x4 28	12) 8 x5 40	13) 5 x3 15	14) 7 x5 35	15) 10 x9 90
16) 5 x4 20	17) 9 x5 45	18) 6 x2 12	19) 6 x7 42	20) 11 x10 110

PAGE: 8

Multiplication Facts

Multiply.

1)	2)	3)	4)	5)
8	11	8	8	9
x4	x0	x6	x2	x1
32	0	48	16	9

6)	7)	8)	9)	10)
3	6	2	2	12
x3	x2	x6	x3	x8
9	12	12	6	96

11)	12)	13)	14)	15)
5	11	12	11	7
x4	x10	x11	x6	x2
20	110	132	66	14

16)	17)	18)	19)	20)
4	12	9	5	11
x4	x12	x8	x7	x5
16	144	72	35	55

PAGE: 9

Multiplication Facts

Multiply.

1)	2)	3)	4)	5)
2	6	10	2	3
x7	x6	x11	x3	x1
14	36	110	6	3

6)	7)	8)	9)	10)
8	4	11	10	2
x7	x3	x5	x9	x5
56	12	55	90	10

11)	12)	13)	14)	15)
12	12	7	12	4
x3	x10	x3	x12	x6
36	120	21	144	24

16)	17)	18)	19)	20)
8	9	9	12	2
x4	x5	x4	x9	x0
32	45	36	108	0

PAGE: 10

Multiplication Facts

Multiply.

1)	2)	3)	4)	5)
9	2	11	11	10
x2	x8	x10	x12	x11
18	16	110	132	110

6)	7)	8)	9)	10)
8	3	10	5	8
x0	x3	x12	x7	x7
0	9	120	35	56

11)	12)	13)	14)	15)
7	2	9	10	10
x2	x6	x5	x7	x3
14	12	45	70	30

16)	17)	18)	19)	20)
9	10	4	11	10
x3	x9	x6	x3	x4
27	90	24	33	40

PAGE: 11

Multiplication Facts

Multiply.

1)	2)	3)	4)	5)
2	3	8	11	12
x4	x0	x4	x5	x11
8	0	32	55	132

6)	7)	8)	9)	10)
2	9	12	10	5
x5	x3	x1	x2	x8
10	27	12	20	40

11)	12)	13)	14)	15)
9	4	10	4	11
x2	x2	x5	x3	x8
18	8	50	12	88

16)	17)	18)	19)	20)
3	6	3	10	12
x6	x9	x2	x8	x6
18	54	6	80	72

PAGE: 12

Multiplication Facts

Multiply.

1)	2)	3)	4)	5)
8	5	6	3	9
x3	x5	x3	x0	x6
24	25	18	0	54

6)	7)	8)	9)	10)
8	5	9	4	11
x1	x8	x7	x9	x7
8	40	63	36	77

11)	12)	13)	14)	15)
3	10	8	11	4
x3	x5	x7	x6	x6
9	50	56	66	24

16)	17)	18)	19)	20)
12	4	10	3	11
x6	x2	x3	x6	x5
72	8	30	18	55

PAGE: 14

Multiplication Find the Missing Multipliers

Fill in the blanks.

1) 7x 3 =21
2) 2x 12 =24
3) 9x 2 =18
4) 10x 2 =20
5) 6x 5 =30
6) 3x 5 =15
7) 4x 3 =12
8) 11x 3 =33
9) 8x 10 =80
10) 10x 9 =90
11) 2x 10 =20
12) 4x 5 =20
13) 10x 0 =0
14) 11x 4 =44
15) 4x 11 =44
16) 7x 8 =56
17) 6x 4 =24
18) 10x 8 =80
19) 11x 9 =99
20) 11x 2 =22

PAGE: 15

Multiplication Find the Missing Multipliers

Fill in the blanks.

1) 12x 4 =48
2) 8x 6 =48
3) 4x 7 =28
4) 8x 10 =80
5) 9x 5 =45
6) 11x 5 =55
7) 3x 5 =15
8) 4x 4 =16
9) 6x 5 =30
10) 8x 4 =32
11) 2x 0 =0
12) 8x 12 =96
13) 2x 5 =10
14) 12x 3 =36
15) 2x 1 =2
16) 2x 2 =4
17) 11x 7 =77
18) 2x 8 =16
19) 12x 9 =108
20) 7x 9 =63

PAGE: 16

Multiplication Find the Missing Multipliers

Fill in the blanks.

1) 10x 11 =110
2) 10x 7 =70
3) 3x 6 =18
4) 10x 3 =30
5) 4x 12 =48
6) 7x 3 =21
7) 11x 2 =22
8) 12x 6 =72
9) 2x 11 =22
10) 8x 5 =40
11) 3x 4 =12
12) 5x 12 =60
13) 6x 5 =30
14) 3x 2 =6
15) 9x 12 =108
16) 10x 1 =10
17) 9x 6 =54
18) 10x 9 =90
19) 8x 0 =0
20) 8x 11 =88

PAGE: 17

Multiplication Find the Missing Multipliers

Fill in the blanks.

1) 8x 4 =32
2) 3x 0 =0
3) 4x 12 =48
4) 4x 3 =12
5) 5x 8 =40
6) 11x 11 =121
7) 7x 7 =49
8) 8x 8 =64
9) 10x 6 =60
10) 11x 1 =11
11) 2x 8 =16
12) 2x 2 =4
13) 6x 11 =66
14) 4x 7 =28
15) 8x 12 =96
16) 5x 12 =60
17) 11x 9 =99
18) 5x 11 =55
19) 12x 5 =60
20) 7x 10 =70

PAGE: 18

Multiplication Find the Missing Multipliers

Fill in the blanks.

1) 8x 11 =88
2) 10x 11 =110
3) 2x 0 =0
4) 6x 8 =48
5) 3x 2 =6
6) 9x 6 =54
7) 2x 6 =12
8) 4x 5 =20
9) 10x 7 =70
10) 2x 9 =18
11) 11x 4 =44
12) 10x 4 =40
13) 9x 9 =81
14) 12x 11 =132
15) 2x 2 =4
16) 2x 1 =2
17) 4x 6 =24
18) 3x 4 =12
19) 10x 5 =50
20) 4x 7 =28

PAGE: 19

Multiplication Find the Missing Multipliers

Fill in the blanks.

1) 3x 4 =12
2) 6x 8 =48
3) 3x 3 =9
4) 2x 0 =0
5) 5x 7 =35
6) 5x 10 =50
7) 9x 11 =99
8) 8x 11 =88
9) 11x 9 =99
10) 6x 5 =30
11) 3x 1 =3
12) 11x 11 =121
13) 2x 12 =24
14) 10x 8 =80
15) 9x 4 =36
16) 8x 9 =72
17) 12x 9 =108
18) 10x 12 =120
19) 9x 5 =45
20) 7x 6 =42

PAGE: 20

Multiplication Find the Missing Multipliers

Fill in the blanks.

1) 6x 1 =6
2) 5x 7 =35
3) 6x 9 =54
4) 2x 3 =6
5) 4x 0 =0
6) 8x 9 =72
7) 3x 10 =30
8) 3x 4 =12
9) 2x 12 =24
10) 2x 7 =14
11) 12x 8 =96
12) 4x 2 =8
13) 10x 6 =60
14) 12x 6 =72
15) 8x 12 =96
16) 7x 4 =28
17) 5x 12 =60
18) 4x 9 =36
19) 6x 11 =66
20) 5x 4 =20

PAGE: 21

Multiplication Find the Missing Multipliers

Fill in the blanks.

1) 10x 1 =10
2) 4x 9 =36
3) 8x 5 =40
4) 4x 5 =20
5) 4x 3 =12
6) 12x 11 =132
7) 11x 8 =88
8) 7x 7 =49
9) 11x 5 =55
10) 3x 11 =33
11) 9x 11 =99
12) 2x 0 =0
13) 3x 3 =9
14) 2x 4 =8
15) 8x 7 =56
16) 7x 5 =35
17) 4x 4 =16
18) 3x 6 =18
19) 10x 11 =110
20) 9x 3 =27

PAGE: 22

Multiplication Find the Missing Multipliers

Fill in the blanks.

1) 7x 11 =77
2) 6x 7 =42
3) 10x 7 =70
4) 3x 12 =36
5) 6x 4 =24
6) 2x 9 =18
7) 3x 10 =30
8) 12x 8 =96
9) 9x 11 =99
10) 9x 6 =54
11) 12x 7 =84
12) 10x 8 =80
13) 12x 12 =144
14) 8x 2 =16
15) 7x 10 =70
16) 11x 7 =77
17) 5x 11 =55
18) 4x 6 =24
19) 4x 11 =44
20) 8x 10 =80

PAGE: 23

Multiplication Find the Missing Multipliers

Fill in the blanks.

1) 11x 2 =22
2) 11x 12 =132
3) 7x 6 =42
4) 2x 5 =10
5) 10x 3 =30
6) 10x 2 =20
7) 11x 10 =110
8) 12x 1 =12
9) 7x 4 =28
10) 7x 0 =0
11) 12x 8 =96
12) 8x 2 =16
13) 4x 10 =40
14) 4x 7 =28
15) 9x 11 =99
16) 6x 8 =48
17) 4x 4 =16
18) 10x 9 =90
19) 6x 7 =42
20) 6x 3 =18

PAGE: 24

Multiplication Find the Missing Multipliers

Fill in the blanks.

1) 2x 8 =16
2) 2x 0 =0
3) 12x 3 =36
4) 6x 9 =54
5) 9x 3 =27
6) 3x 3 =9
7) 2x 2 =4
8) 5x 2 =10
9) 2x 6 =12
10) 6x 2 =12
11) 3x 9 =27
12) 4x 5 =20
13) 9x 12 =108
14) 11x 5 =55
15) 7x 1 =7
16) 5x 12 =60
17) 9x 4 =36
18) 10x 5 =50
19) 3x 12 =36
20) 9x 5 =45

PAGE: 26

Division Facts

Divide.

1) 63÷7= 9
2) 0÷5= 0
3) 120÷10= 12
4) 48÷8= 6
5) 6÷3= 2
6) 66÷6= 11
7) 18÷9= 2
8) 60÷10= 6
9) 7÷1= 7
10) 24÷6= 4
11) 30÷5= 6
12) 72÷6= 12
13) 22÷11= 2
14) 22÷2= 11
15) 12÷6= 2
16) 40÷4= 10
17) 99÷9= 11
18) 80÷10= 8
19) 36÷12= 3
20) 49÷7= 7

PAGE: 27

Division Facts

Divide.

1) 64÷8= 8
2) 72÷6= 12
3) 0÷6= 0
4) 36÷12= 3
5) 88÷8= 11
6) 24÷8= 3
7) 9÷9= 1
8) 84÷7= 12
9) 20÷5= 4
10) 2÷1= 2
11) 50÷5= 10
12) 48÷4= 12
13) 12÷6= 2
14) 84÷12= 7
15) 96÷12= 8
16) 24÷6= 4
17) 30÷6= 5
18) 24÷12= 2
19) 18÷3= 6
20) 18÷9= 2

PAGE: 28

Division Facts

Divide.

1) 0÷4=	0	11) 90÷9=	10
2) 50÷10=	5	12) 15÷3=	5
3) 8÷4=	2	13) 11÷1=	11
4) 18÷9=	2	14) 108÷9=	12
5) 44÷4=	11	15) 84÷12=	7
6) 24÷6=	4	16) 60÷12=	5
7) 21÷3=	7	17) 12÷3=	4
8) 63÷7=	9	18) 15÷5=	3
9) 63÷9=	7	19) 42÷6=	7
10) 110÷10=	11	20) 120÷10=	12

PAGE: 29

Division Facts

Divide.

1) 30÷3=	10	11) 18÷6=	3
2) 0÷2=	0	12) 12÷2=	6
3) 27÷9=	3	13) 132÷11=	12
4) 27÷3=	9	14) 48÷8=	6
5) 2÷2=	1	15) 49÷7=	7
6) 42÷6=	7	16) 24÷8=	3
7) 144÷12=	12	17) 40÷8=	5
8) 9÷1=	9	18) 16÷2=	8
9) 77÷7=	11	19) 66÷6=	11
10) 42÷7=	6	20) 44÷4=	11

PAGE: 30

Division Facts

Divide.

1) 56÷7=	8	11) 30÷3=	10
2) 40÷4=	10	12) 50÷5=	10
3) 132÷12=	11	13) 10÷1=	10
4) 72÷12=	6	14) 7÷7=	1
5) 90÷10=	9	15) 16÷2=	8
6) 24÷12=	2	16) 24÷8=	3
7) 45÷5=	9	17) 48÷6=	8
8) 72÷6=	12	18) 14÷7=	2
9) 12÷3=	4	19) 15÷5=	3
10) 8÷4=	2	20) 12÷6=	2

PAGE: 31

Division Facts

Divide.

1) 30÷3=	10	11) 9÷1=	9
2) 44÷4=	11	12) 16÷4=	4
3) 0÷12=	0	13) 60÷5=	12
4) 12÷3=	4	14) 2÷2=	1
5) 72÷12=	6	15) 25÷5=	5
6) 24÷6=	4	16) 36÷4=	9
7) 6÷2=	3	17) 28÷7=	4
8) 32÷4=	8	18) 80÷10=	8
9) 49÷7=	7	19) 20÷10=	2
10) 77÷7=	11	20) 42÷7=	6

PAGE: 32

Division Facts

Divide.

1) 48÷4=	12	11) 0÷5=	0
2) 132÷12=	11	12) 40÷10=	4
3) 132÷11=	12	13) 10÷5=	2
4) 36÷4=	9	14) 3÷3=	1
5) 18÷2=	9	15) 84÷12=	7
6) 3÷1=	3	16) 18÷6=	3
7) 6÷6=	1	17) 80÷8=	10
8) 63÷7=	9	18) 30÷10=	3
9) 108÷12=	9	19) 48÷8=	6
10) 42÷6=	7	20) 49÷7=	7

PAGE: 33

Division Facts

Divide.

1) 9÷1=	9	11) 18÷6=	3
2) 21÷7=	3	12) 77÷7=	11
3) 40÷10=	4	13) 32÷8=	4
4) 24÷12=	2	14) 132÷12=	11
5) 0÷5=	0	15) 28÷7=	4
6) 2÷2=	1	16) 54÷6=	9
7) 63÷7=	9	17) 60÷6=	10
8) 24÷8=	3	18) 48÷8=	6
9) 10÷10=	1	19) 30÷10=	3
10) 20÷5=	4	20) 40÷8=	5

PAGE: 34

Division Facts

Divide.

1) 70÷10= 7
2) 35÷7= 5
3) 80÷8= 10
4) 54÷9= 6
5) 84÷12= 7
6) 100÷10= 10
7) 3÷1= 3
8) 0÷3= 0
9) 54÷6= 9
10) 22÷11= 2
11) 99÷9= 11
12) 108÷9= 12
13) 120÷10= 12
14) 10÷2= 5
15) 16÷8= 2
16) 30÷5= 6
17) 8÷4= 2
18) 24÷8= 3
19) 9÷9= 1
20) 16÷2= 8

PAGE: 35

Division Facts

Divide.

1) 110÷11= 10
2) 35÷5= 7
3) 6÷2= 3
4) 44÷11= 4
5) 84÷12= 7
6) 0÷6= 0
7) 96÷12= 8
8) 36÷3= 12
9) 84÷7= 12
10) 48÷4= 12
11) 16÷2= 8
12) 1÷1= 1
13) 63÷9= 7
14) 24÷3= 8
15) 16÷4= 4
16) 18÷2= 9
17) 9÷3= 3
18) 15÷5= 3
19) 10÷10= 1
20) 99÷11= 9

PAGE: 36

Division Facts

Divide.

1) 28÷4= 7
2) 9÷1= 9
3) 22÷11= 2
4) 48÷4= 12
5) 10÷5= 2
6) 0÷10= 0
7) 40÷4= 10
8) 14÷7= 2
9) 60÷10= 6
10) 63÷7= 9
11) 56÷7= 8
12) 144÷12= 12
13) 48÷6= 8
14) 33÷11= 3
15) 90÷9= 10
16) 20÷4= 5
17) 16÷2= 8
18) 6÷6= 1
19) 44÷11= 4
20) 14÷2= 7

PAGE: 38

Division Find the Missing Divisors

Fill in the blanks.

1) 3÷ 3 =1
2) 1÷ 1 =1
3) 30÷ 10 =3
4) 18÷ 3 =6
5) 12÷ 4 =3
6) 54÷ 9 =6
7) 48÷ 6 =8
8) 20÷ 2 =10
9) 60÷ 12 =5
10) 24÷ 6 =4
11) 28÷ 4 =7
12) 40÷ 5 =8
13) 12÷ 2 =6
14) 80÷ 10 =8
15) 16÷ 4 =4
16) 48÷ 12 =4
17) 88÷ 8 =11
18) 56÷ 7 =8
19) 90÷ 9 =10
20) 15÷ 5 =3

PAGE: 39

Division Find the Missing Divisors

Fill in the blanks.

1) 108÷ 12 =9
2) 18÷ 9 =2
3) 96÷ 12 =8
4) 11÷ 1 =11
5) 16÷ 4 =4
6) 20÷ 5 =4
7) 32÷ 4 =8
8) 18÷ 2 =9
9) 49÷ 7 =7
10) 45÷ 9 =5
11) 24÷ 3 =8
12) 10÷ 10 =1
13) 121÷ 11 =11
14) 63÷ 7 =9
15) 84÷ 7 =12
16) 16÷ 2 =8
17) 12÷ 12 =1
18) 33÷ 11 =3
19) 33÷ 3 =11
20) 100÷ 10 =10

PAGE: 40

Division Find the Missing Divisors

Fill in the blanks.

1) 50÷ 10 =5
2) 5÷ 5 =1
3) 16÷ 8 =2
4) 70÷ 7 =10
5) 66÷ 6 =11
6) 14÷ 2 =7
7) 84÷ 12 =7
8) 120÷ 10 =12
9) 108÷ 12 =9
10) 99÷ 9 =11
11) 40÷ 5 =8
12) 20÷ 10 =2
13) 40÷ 4 =10
14) 25÷ 5 =5
15) 16÷ 2 =8
16) 12÷ 3 =4
17) 96÷ 8 =12
18) 7÷ 1 =7
19) 49÷ 7 =7
20) 4÷ 2 =2

PAGE: 41

Division Find the Missing Divisors

Fill in the blanks.

1) 8÷ 8 =1
2) 22÷ 11 =2
3) 35÷ 7 =5
4) 48÷ 4 =12
5) 6÷ 2 =3
6) 18÷ 9 =2
7) 56÷ 8 =7
8) 1÷ 1 =1
9) 63÷ 7 =9
10) 28÷ 7 =4
11) 55÷ 5 =11
12) 32÷ 4 =8
13) 100÷ 10 =10
14) 36÷ 3 =12
15) 24÷ 4 =6
16) 77÷ 7 =11
17) 72÷ 12 =6
18) 16÷ 4 =4
19) 72÷ 9 =8
20) 5÷ 5 =1

PAGE: 42

Division Find the Missing Divisors

Fill in the blanks.

1) 120÷ 10 =12
2) 45÷ 9 =5
3) 20÷ 2 =10
4) 60÷ 5 =12
5) 88÷ 8 =11
6) 25÷ 5 =5
7) 12÷ 12 =1
8) 36÷ 3 =12
9) 63÷ 9 =7
10) 99÷ 9 =11
11) 54÷ 9 =6
12) 4÷ 1 =4
13) 80÷ 10 =8
14) 5÷ 5 =1
15) 44÷ 4 =11
16) 14÷ 2 =7
17) 42÷ 6 =7
18) 77÷ 7 =11
19) 16÷ 4 =4
20) 24÷ 6 =4

PAGE: 43

Division Find the Missing Divisors

Fill in the blanks.

1) 42÷ 6 =7
2) 84÷ 12 =7
3) 3÷ 1 =3
4) 6÷ 2 =3
5) 20÷ 10 =2
6) 33÷ 11 =3
7) 45÷ 5 =9
8) 110÷ 11 =10
9) 16÷ 8 =2
10) 49÷ 7 =7
11) 24÷ 8 =3
12) 40÷ 10 =4
13) 35÷ 7 =5
14) 120÷ 12 =10
15) 108÷ 12 =9
16) 36÷ 12 =3
17) 24÷ 3 =8
18) 12÷ 12 =1
19) 36÷ 4 =9
20) 36÷ 6 =6

PAGE: 44

Division Find the Missing Divisors

Fill in the blanks.

1) 44÷ 11 =4
2) 20÷ 5 =4
3) 132÷ 11 =12
4) 10÷ 2 =5
5) 44÷ 4 =11
6) 70÷ 10 =7
7) 100÷ 10 =10
8) 88÷ 11 =8
9) 18÷ 3 =6
10) 16÷ 4 =4
11) 14÷ 2 =7
12) 81÷ 9 =9
13) 12÷ 12 =1
14) 30÷ 5 =6
15) 108÷ 9 =12
16) 24÷ 4 =6
17) 2÷ 1 =2
18) 24÷ 3 =8
19) 24÷ 8 =3
20) 32÷ 4 =8

PAGE: 45

Division Find the Missing Divisors

Fill in the blanks.

1) 55÷ 5 =11
2) 10÷ 10 =1
3) 11÷ 11 =1
4) 10÷ 1 =10
5) 56÷ 8 =7
6) 66÷ 11 =6
7) 24÷ 8 =3
8) 16÷ 2 =8
9) 84÷ 7 =12
10) 25÷ 5 =5
11) 45÷ 9 =5
12) 108÷ 9 =12
13) 77÷ 11 =7
14) 20÷ 4 =5
15) 32÷ 4 =8
16) 63÷ 7 =9
17) 12÷ 3 =4
18) 42÷ 7 =6
19) 72÷ 6 =12
20) 36÷ 3 =12

PAGE: 46

Division Find the Missing Divisors

Fill in the blanks.

1) 40÷ 4 =10
2) 45÷ 9 =5
3) 55÷ 11 =5
4) 48÷ 4 =12
5) 12÷ 2 =6
6) 132÷ 11 =12
7) 100÷ 10 =10
8) 4÷ 4 =1
9) 14÷ 7 =2
10) 3÷ 3 =1
11) 42÷ 6 =7
12) 63÷ 9 =7
13) 10÷ 10 =1
14) 11÷ 1 =11
15) 9÷ 3 =3
16) 90÷ 9 =10
17) 60÷ 6 =10
18) 9÷ 9 =1
19) 32÷ 4 =8
20) 32÷ 8 =4

PAGE: 47

Division Find the Missing Divisors

Fill in the blanks.

1) 12÷ 4 =3
2) 40÷ 5 =8
3) 72÷ 8 =9
4) 49÷ 7 =7
5) 35÷ 5 =7
6) 56÷ 7 =8
7) 16÷ 4 =4
8) 7÷ 1 =7
9) 30÷ 6 =5
10) 90÷ 10 =9
11) 24÷ 6 =4
12) 15÷ 3 =5
13) 81÷ 9 =9
14) 48÷ 4 =12
15) 60÷ 10 =6
16) 84÷ 7 =12
17) 36÷ 6 =6
18) 110÷ 10 =11
19) 20÷ 10 =2
20) 66÷ 6 =11

PAGE: 48

Division Find the Missing Divisors

Fill in the blanks.

1) 70÷ 7 =10
2) 48÷ 12 =4
3) 54÷ 6 =9
4) 24÷ 4 =6
5) 72÷ 12 =6
6) 28÷ 7 =4
7) 99÷ 9 =11
8) 48÷ 6 =8
9) 35÷ 5 =7
10) 96÷ 8 =12
11) 14÷ 7 =2
12) 1÷ 1 =1
13) 56÷ 7 =8
14) 24÷ 12 =2
15) 21÷ 7 =3
16) 6÷ 3 =2
17) 24÷ 3 =8
18) 66÷ 6 =11
19) 40÷ 10 =4
20) 84÷ 12 =7

PAGE: 50

Multiplication 2-Digit Multiplicands x 1-Digit Multipliers

Multiply.

1) 84 x9 756	2) 97 x3 291	3) 11 x2 22	4) 58 x6 348	5) 26 x6 156
6) 50 x3 150	7) 47 x5 235	8) 58 x4 232	9) 38 x6 228	10) 22 x4 88
11) 97 x1 97	12) 20 x2 40	13) 97 x6 582	14) 81 x2 162	15) 41 x5 205
16) 91 x4 364	17) 24 x5 120	18) 56 x6 336	19) 32 x2 64	20) 86 x8 688

PAGE: 51

Multiplication 2-Digit Multiplicands x 1-Digit Multipliers

Multiply.

1) 74 x2 148	2) 40 x6 240	3) 67 x1 67	4) 81 x7 567	5) 17 x6 102
6) 25 x9 225	7) 75 x6 450	8) 52 x7 364	9) 46 x8 368	10) 19 x2 38
11) 55 x6 330	12) 44 x5 220	13) 26 x6 156	14) 87 x5 435	15) 39 x5 195
16) 38 x9 342	17) 24 x9 216	18) 60 x9 540	19) 35 x7 245	20) 96 x9 864

PAGE: 52

Multiplication 2-Digit Multiplicands x 1-Digit Multipliers

Multiply.

1) 35 x4 140	2) 52 x4 208	3) 14 x9 126	4) 21 x3 63	5) 49 x3 147
6) 16 x4 64	7) 15 x2 30	8) 56 x2 112	9) 99 x7 693	10) 48 x5 240
11) 86 x6 516	12) 20 x9 180	13) 73 x3 219	14) 47 x5 235	15) 24 x2 48
16) 70 x6 420	17) 37 x7 259	18) 76 x2 152	19) 76 x6 456	20) 95 x7 665

PAGE: 53

Multiplication 2-Digit Multiplicands x 1-Digit Multipliers

Multiply.

1) 29 x8 232	2) 94 x8 752	3) 39 x3 117	4) 19 x6 114	5) 66 x8 528
6) 61 x5 305	7) 74 x2 148	8) 72 x8 576	9) 72 x6 432	10) 36 x5 180
11) 22 x7 154	12) 76 x2 152	13) 24 x7 168	14) 86 x7 602	15) 40 x5 200
16) 98 x5 490	17) 79 x1 79	18) 99 x7 693	19) 14 x3 42	20) 28 x7 196

PAGE: 54

Multiplication 2-Digit Multiplicands x 1-Digit Multipliers

Multiply.

1) 64 x5 320	2) 83 x6 498	3) 49 x7 343	4) 87 x3 261	5) 32 x4 128
6) 34 x7 238	7) 97 x2 194	8) 44 x5 220	9) 41 x3 123	10) 15 x5 75
11) 79 x4 316	12) 66 x2 132	13) 39 x7 273	14) 93 x5 465	15) 64 x2 128
16) 39 x1 39	17) 72 x7 504	18) 67 x7 469	19) 28 x2 56	20) 31 x9 279

PAGE: 55

Multiplication 2-Digit Multiplicands x 1-Digit Multipliers

Multiply.

1) 73 x7 511	2) 39 x2 78	3) 67 x6 402	4) 66 x5 330	5) 90 x2 180
6) 42 x4 168	7) 14 x2 28	8) 86 x5 430	9) 30 x8 240	10) 97 x6 582
11) 99 x9 891	12) 13 x2 26	13) 50 x5 250	14) 92 x3 276	15) 28 x4 112
16) 94 x4 376	17) 39 x7 273	18) 95 x4 380	19) 49 x4 196	20) 36 x8 288

PAGE: 56

Multiplication 2-Digit Multiplicands x 1-Digit Multipliers

Multiply.

1) 54 x8 432	2) 24 x9 216	3) 52 x3 156	4) 79 x6 474	5) 77 x6 462
6) 26 x1 26	7) 71 x5 355	8) 51 x5 255	9) 18 x8 144	10) 30 x6 180
11) 59 x7 413	12) 13 x5 65	13) 30 x5 150	14) 81 x2 162	15) 25 x3 75
16) 44 x5 220	17) 16 x7 112	18) 93 x6 558	19) 16 x3 48	20) 65 x5 325

PAGE: 57

Multiplication 2-Digit Multiplicands x 1-Digit Multipliers

Multiply.

1) 10 x5 50	2) 31 x1 31	3) 23 x8 184	4) 18 x3 54	5) 14 x5 70
6) 38 x3 114	7) 37 x4 148	8) 32 x6 192	9) 99 x5 495	10) 75 x2 150
11) 75 x7 525	12) 20 x4 80	13) 63 x2 126	14) 31 x3 93	15) 26 x6 156
16) 49 x4 196	17) 61 x6 366	18) 46 x7 322	19) 97 x5 485	20) 75 x4 300

PAGE: 58

Multiplication 2-Digit Multiplicands x 1-Digit Multipliers

Multiply.

1) 19 x7 133	2) 98 x1 98	3) 67 x9 603	4) 77 x8 616	5) 80 x7 560
6) 35 x4 140	7) 33 x7 231	8) 63 x9 567	9) 76 x5 380	10) 56 x6 336
11) 84 x6 504	12) 46 x8 368	13) 63 x3 189	14) 18 x5 90	15) 69 x2 138
16) 93 x2 186	17) 18 x6 108	18) 77 x7 539	19) 56 x4 224	20) 91 x7 637

PAGE: 59

Multiplication 2-Digit Multiplicands x 1-Digit Multipliers

Multiply.

1) 94 x7 658	2) 82 x5 410	3) 87 x2 174	4) 71 x2 142	5) 28 x3 84
6) 52 x4 208	7) 39 x4 156	8) 33 x2 66	9) 16 x6 96	10) 68 x3 204
11) 78 x4 312	12) 76 x9 684	13) 33 x6 198	14) 20 x1 20	15) 91 x8 728
16) 88 x5 440	17) 59 x9 531	18) 63 x3 189	19) 56 x5 280	20) 45 x5 225

PAGE: 60

Multiplication 2-Digit Multiplicands x 1-Digit Multipliers

Multiply.

#	Multiplicand	Multiplier	Answer
1)	19	x9	171
2)	18	x6	108
3)	68	x9	612
4)	10	x3	30
5)	72	x4	288
6)	44	x5	220
7)	61	x7	427
8)	40	x7	280
9)	21	x9	189
10)	51	x7	357
11)	73	x5	365
12)	84	x5	420
13)	22	x3	66
14)	14	x8	112
15)	81	x9	729
16)	29	x2	58
17)	54	x4	216
18)	92	x9	828
19)	63	x9	567
20)	32	x4	128

PAGE: 62

Multiplication 4-Digit Multiplicands x 1-Digit Multipliers

Multiply.

#	Multiplicand	Multiplier	Answer
1)	2159	x6	12954
2)	1826	x3	5478
3)	4900	x4	19600
4)	3328	x5	16640
5)	6780	x7	47460
6)	4391	x5	21955
7)	7517	x3	22551
8)	2482	x4	9928
9)	2330	x9	20970
10)	8572	x3	25716
11)	8938	x6	53628
12)	2129	x3	6387
13)	4560	x2	9120
14)	3121	x4	12484
15)	6915	x3	20745

PAGE: 63

Multiplication 4-Digit Multiplicands x 1-Digit Multipliers

Multiply.

#	Multiplicand	Multiplier	Answer
1)	9017	x7	63119
2)	2220	x8	17760
3)	3786	x5	18930
4)	8196	x8	65568
5)	2898	x8	23184
6)	2060	x2	4120
7)	8319	x5	41595
8)	1532	x3	4596
9)	8520	x9	76680
10)	1829	x4	7316
11)	5470	x2	10940
12)	2672	x2	5344
13)	5635	x3	16905
14)	8278	x6	49668
15)	1987	x2	3974

PAGE: 64

Multiplication 4-Digit Multiplicands x 1-Digit Multipliers

Multiply.

#	Multiplicand	Multiplier	Answer
1)	8728	x3	26184
2)	1694	x9	15246
3)	9843	x7	68901
4)	7968	x5	39840
5)	4331	x2	8662
6)	2413	x7	16891
7)	7661	x5	38305
8)	7246	x8	57968
9)	3802	x8	30416
10)	4186	x4	16744
11)	8222	x4	32888
12)	6452	x4	25808
13)	4889	x2	9778
14)	4764	x5	23820
15)	9274	x3	27822

PAGE: 65

Multiplication 4-Digit Multiplicands x 1-Digit Multipliers

Multiply.

#	Multiplicand	Multiplier	Answer
1)	3339	x8	26712
2)	5344	x7	37408
3)	1896	x8	15168
4)	8569	x8	68552
5)	4655	x8	37240
6)	2916	x5	14580
7)	2334	x4	9336
8)	3856	x8	30848
9)	8151	x4	32604
10)	4356	x9	39204
11)	3825	x5	19125
12)	1885	x2	3770
13)	2576	x9	23184
14)	7412	x2	14824
15)	4122	x3	12366

PAGE: 66

Multiplication 4-Digit Multiplicands x 1-Digit Multipliers

Multiply.

#	Multiplicand	Multiplier	Answer
1)	9406	x4	37624
2)	3682	x5	18410
3)	9836	x6	59016
4)	9973	x3	29919
5)	6169	x4	24676
6)	9216	x6	55296
7)	9376	x3	28128
8)	1129	x3	3387
9)	7925	x7	55475
10)	7931	x2	15862
11)	3784	x6	22704
12)	2340	x9	21060
13)	3423	x3	10269
14)	6575	x8	52600
15)	8809	x6	52854

PAGE: 67

Multiplication 4-Digit Multiplicands x 1-Digit Multipliers

Multiply.

1) 2355 x4 9420	2) 8413 x8 67304	3) 2899 x5 14495
4) 4202 x8 33616	5) 7926 x8 63408	6) 2645 x2 5290
7) 3653 x7 25571	8) 5711 x4 22844	9) 7800 x5 39000
10) 1869 x8 14952	11) 6586 x4 26344	12) 2088 x6 12528
13) 8730 x8 69840	14) 3836 x7 26852	15) 9449 x3 28347

PAGE: 68

Multiplication 4-Digit Multiplicands x 1-Digit Multipliers

Multiply.

1) 2198 x8 17584	2) 8229 x6 49374	3) 4447 x6 26682
4) 3051 x8 24408	5) 6960 x6 41760	6) 7559 x6 45354
7) 9455 x5 47275	8) 1221 x8 9768	9) 7151 x2 14302
10) 4006 x6 24036	11) 8767 x3 26301	12) 4383 x3 13149
13) 9522 x9 85698	14) 4320 x6 25920	15) 9134 x8 73072

PAGE: 69

Multiplication 4-Digit Multiplicands x 1-Digit Multipliers

Multiply.

1) 1287 x9 11583	2) 2822 x9 25398	3) 6252 x2 12504
4) 8306 x8 66448	5) 7683 x5 38415	6) 2803 x8 22424
7) 7550 x9 67950	8) 4032 x3 12096	9) 3322 x8 26576
10) 7213 x4 28852	11) 5966 x8 47728	12) 3063 x6 18378
13) 3952 x6 23712	14) 4802 x3 14406	15) 8249 x7 57743

PAGE: 70

Multiplication 4-Digit Multiplicands x 1-Digit Multipliers

Multiply.

1) 3149 x8 25192	2) 7325 x6 43950	3) 2321 x4 9284
4) 3149 x9 28341	5) 2039 x5 10195	6) 2190 x5 10950
7) 6520 x4 26080	8) 6417 x7 44919	9) 6484 x5 32420
10) 4297 x2 8594	11) 6149 x9 55341	12) 6801 x5 34005
13) 5234 x5 26170	14) 2397 x3 7191	15) 7957 x9 71613

PAGE: 71

Multiplication 4-Digit Multiplicands x 1-Digit Multipliers

Multiply.

1) 4910 x7 34370	2) 1893 x5 9465	3) 4285 x2 8570
4) 5443 x2 10886	5) 9689 x7 67823	6) 3069 x8 24552
7) 5092 x2 10184	8) 3489 x6 20934	9) 5961 x6 35766
10) 1924 x2 3848	11) 5265 x3 15795	12) 7091 x4 28364
13) 9474 x4 37896	14) 5209 x2 10418	15) 2556 x6 15336

PAGE: 72

Multiplication 4-Digit Multiplicands x 1-Digit Multipliers

Multiply.

1) 7355 x9 66195	2) 8587 x5 42935	3) 4928 x2 9856
4) 7453 x8 59624	5) 8953 x4 35812	6) 8691 x8 69528
7) 1147 x5 5735	8) 9771 x9 87939	9) 3730 x3 11190
10) 3661 x2 7322	11) 7420 x5 37100	12) 5831 x5 29155
13) 7437 x4 29748	14) 9249 x8 73992	15) 3498 x2 6996

PAGE: 74

Multiplication 3-Digit Multiplicands x 2-Digit Multipliers

Multiply.

1) 120 x79 = 9480	2) 873 x34 = 29682	3) 840 x82 = 68880
4) 833 x75 = 62475	5) 623 x43 = 26789	6) 517 x78 = 40326
7) 229 x48 = 10992	8) 459 x18 = 8262	9) 866 x50 = 43300

PAGE: 75

Multiplication 3-Digit Multiplicands x 2-Digit Multipliers

Multiply.

1) 811 x43 = 34873	2) 837 x58 = 48546	3) 263 x37 = 9731
4) 380 x93 = 35340	5) 761 x91 = 69251	6) 841 x70 = 58870
7) 589 x45 = 26505	8) 282 x74 = 20868	9) 282 x53 = 14946

PAGE: 76

Multiplication 3-Digit Multiplicands x 2-Digit Multipliers

Multiply.

1) 307 x86 = 26402	2) 865 x74 = 64010	3) 145 x10 = 1450
4) 871 x73 = 63583	5) 227 x78 = 17706	6) 432 x85 = 36720
7) 413 x16 = 6608	8) 982 x84 = 82488	9) 112 x47 = 5264

PAGE: 77

Multiplication 3-Digit Multiplicands x 2-Digit Multipliers

Multiply.

1) 572 x36 = 20592	2) 879 x74 = 65046	3) 765 x51 = 39015
4) 368 x20 = 7360	5) 583 x34 = 19822	6) 143 x29 = 4147
7) 787 x74 = 58238	8) 739 x76 = 56164	9) 784 x32 = 25088

PAGE: 78

Multiplication 3-Digit Multiplicands x 2-Digit Multipliers

Multiply.

1) 729 x16 = 11664	2) 219 x37 = 8103	3) 951 x32 = 30432
4) 503 x58 = 29174	5) 138 x12 = 1656	6) 976 x59 = 57584
7) 173 x67 = 11591	8) 595 x38 = 22610	9) 692 x81 = 56052

PAGE: 79

Multiplication 3-Digit Multiplicands x 2-Digit Multipliers

Multiply.

1) 770 x29 = 22330	2) 978 x92 = 89976	3) 157 x34 = 5338
4) 892 x31 = 27652	5) 899 x52 = 46748	6) 909 x19 = 17271
7) 831 x56 = 46536	8) 439 x11 = 4829	9) 937 x67 = 62779

PAGE: 80

Multiplication 3-Digit Multiplicands x 2-Digit Multipliers

Multiply.

1) 441 x91 = 40131
2) 436 x85 = 37060
3) 682 x47 = 32054
4) 950 x96 = 91200
5) 647 x56 = 36232
6) 387 x62 = 23994
7) 870 x91 = 79170
8) 856 x25 = 21400
9) 537 x82 = 44034

PAGE: 81

Multiplication 3-Digit Multiplicands x 2-Digit Multipliers

Multiply.

1) 239 x41 = 9799
2) 238 x56 = 13328
3) 225 x45 = 10125
4) 448 x28 = 12544
5) 117 x23 = 2691
6) 775 x23 = 17825
7) 675 x63 = 42525
8) 854 x17 = 14518
9) 409 x27 = 11043

PAGE: 82

Multiplication 3-Digit Multiplicands x 2-Digit Multipliers

Multiply.

1) 254 x89 = 22606
2) 828 x52 = 43056
3) 240 x58 = 13920
4) 680 x40 = 27200
5) 823 x41 = 33743
6) 101 x13 = 1313
7) 285 x60 = 17100
8) 751 x60 = 45060
9) 962 x88 = 84656

PAGE: 83

Multiplication 3-Digit Multiplicands x 2-Digit Multipliers

Multiply.

1) 931 x88 = 81928
2) 531 x20 = 10620
3) 377 x60 = 22620
4) 777 x34 = 26418
5) 545 x77 = 41965
6) 382 x74 = 28268
7) 845 x90 = 76050
8) 329 x71 = 23359
9) 237 x98 = 23226

PAGE: 84

Multiplication 3-Digit Multiplicands x 2-Digit Multipliers

Multiply.

1) 744 x93 = 69192
2) 595 x47 = 27965
3) 106 x10 = 1060
4) 625 x79 = 49375
5) 430 x57 = 24510
6) 814 x56 = 45584
7) 450 x29 = 13050
8) 572 x18 = 10296
9) 250 x37 = 9250

PAGE: 86

Division 3-Digit Dividends / 1-Digit Divisors

Divide.

1) 4)192 = 48
2) 4)348 = 87
3) 5)115 = 23
4) 6)150 = 25
5) 3)108 = 36
6) 9)522 = 58
7) 3)147 = 49
8) 2)146 = 73
9) 9)144 = 16
10) 2)180 = 90
11) 6)492 = 82
12) 7)497 = 71

PAGE: 87

Division 3-Digit Dividends / 1-Digit Divisors

Divide.

1) 8)720 = 90	2) 9)270 = 30	3) 5)340 = 68
4) 2)120 = 60	5) 7)497 = 71	6) 8)664 = 83
7) 2)116 = 58	8) 7)588 = 84	9) 8)752 = 94
10) 8)136 = 17	11) 4)212 = 53	12) 2)156 = 78

PAGE: 88

Division 3-Digit Dividends / 1-Digit Divisors

Divide.

1) 7)343 = 49	2) 4)200 = 50	3) 4)196 = 49
4) 2)170 = 85	5) 7)588 = 84	6) 4)160 = 40
7) 2)114 = 57	8) 5)125 = 25	9) 6)198 = 33
10) 8)584 = 73	11) 9)234 = 26	12) 9)756 = 84

PAGE: 89

Division 3-Digit Dividends / 1-Digit Divisors

Divide.

1) 4)336 = 84	2) 6)528 = 88	3) 5)155 = 31
4) 4)280 = 70	5) 4)216 = 54	6) 5)280 = 56
7) 9)117 = 13	8) 5)270 = 54	9) 9)675 = 75
10) 5)180 = 36	11) 9)234 = 26	12) 9)594 = 66

PAGE: 90

Division 3-Digit Dividends / 1-Digit Divisors

Divide.

1) 5)335 = 67	2) 5)345 = 69	3) 3)282 = 94
4) 6)192 = 32	5) 3)120 = 40	6) 3)147 = 49
7) 7)476 = 68	8) 4)192 = 48	9) 8)112 = 14
10) 6)258 = 43	11) 8)696 = 87	12) 6)438 = 73

PAGE: 91

Division 3-Digit Dividends / 1-Digit Divisors

Divide.

1) 3)168 = 56	2) 9)207 = 23	3) 4)196 = 49
4) 8)728 = 91	5) 8)504 = 63	6) 7)518 = 74
7) 4)212 = 53	8) 7)567 = 81	9) 5)430 = 86
10) 9)711 = 79	11) 9)189 = 21	12) 3)198 = 66

PAGE: 92

Division 3-Digit Dividends / 1-Digit Divisors

Divide.

1) 8)288 = 36	2) 7)112 = 16	3) 4)256 = 64
4) 8)640 = 80	5) 9)189 = 21	6) 9)585 = 65
7) 6)138 = 23	8) 5)295 = 59	9) 5)465 = 93
10) 8)352 = 44	11) 8)560 = 70	12) 9)162 = 18

PAGE: 93

Division 3-Digit Dividends / 1-Digit Divisors

Divide.

1) $\begin{array}{r}22\\9\overline{)198}\end{array}$	2) $\begin{array}{r}45\\4\overline{)180}\end{array}$	3) $\begin{array}{r}54\\9\overline{)486}\end{array}$
4) $\begin{array}{r}66\\2\overline{)132}\end{array}$	5) $\begin{array}{r}56\\4\overline{)224}\end{array}$	6) $\begin{array}{r}38\\9\overline{)342}\end{array}$
7) $\begin{array}{r}65\\9\overline{)585}\end{array}$	8) $\begin{array}{r}42\\8\overline{)336}\end{array}$	9) $\begin{array}{r}91\\7\overline{)637}\end{array}$
10) $\begin{array}{r}88\\8\overline{)704}\end{array}$	11) $\begin{array}{r}86\\3\overline{)258}\end{array}$	12) $\begin{array}{r}48\\9\overline{)432}\end{array}$

PAGE: 94

Division 3-Digit Dividends / 1-Digit Divisors

Divide.

1) $\begin{array}{r}77\\5\overline{)385}\end{array}$	2) $\begin{array}{r}18\\6\overline{)108}\end{array}$	3) $\begin{array}{r}15\\8\overline{)120}\end{array}$
4) $\begin{array}{r}81\\8\overline{)648}\end{array}$	5) $\begin{array}{r}71\\7\overline{)497}\end{array}$	6) $\begin{array}{r}91\\4\overline{)364}\end{array}$
7) $\begin{array}{r}80\\2\overline{)160}\end{array}$	8) $\begin{array}{r}48\\3\overline{)144}\end{array}$	9) $\begin{array}{r}55\\3\overline{)165}\end{array}$
10) $\begin{array}{r}50\\5\overline{)250}\end{array}$	11) $\begin{array}{r}73\\7\overline{)511}\end{array}$	12) $\begin{array}{r}29\\8\overline{)232}\end{array}$

PAGE: 95

Division 3-Digit Dividends / 1-Digit Divisors

Divide.

1) $\begin{array}{r}48\\5\overline{)240}\end{array}$	2) $\begin{array}{r}34\\4\overline{)136}\end{array}$	3) $\begin{array}{r}47\\8\overline{)376}\end{array}$
4) $\begin{array}{r}49\\8\overline{)392}\end{array}$	5) $\begin{array}{r}64\\2\overline{)128}\end{array}$	6) $\begin{array}{r}74\\4\overline{)296}\end{array}$
7) $\begin{array}{r}75\\7\overline{)525}\end{array}$	8) $\begin{array}{r}63\\9\overline{)567}\end{array}$	9) $\begin{array}{r}97\\8\overline{)776}\end{array}$
10) $\begin{array}{r}13\\8\overline{)104}\end{array}$	11) $\begin{array}{r}54\\2\overline{)108}\end{array}$	12) $\begin{array}{r}35\\5\overline{)175}\end{array}$

PAGE: 96

Division 3-Digit Dividends / 1-Digit Divisors

Divide.

1) $\begin{array}{r}95\\7\overline{)665}\end{array}$	2) $\begin{array}{r}34\\7\overline{)238}\end{array}$	3) $\begin{array}{r}89\\9\overline{)801}\end{array}$
4) $\begin{array}{r}34\\8\overline{)272}\end{array}$	5) $\begin{array}{r}33\\7\overline{)231}\end{array}$	6) $\begin{array}{r}98\\4\overline{)392}\end{array}$
7) $\begin{array}{r}57\\3\overline{)171}\end{array}$	8) $\begin{array}{r}70\\5\overline{)350}\end{array}$	9) $\begin{array}{r}13\\8\overline{)104}\end{array}$
10) $\begin{array}{r}25\\7\overline{)175}\end{array}$	11) $\begin{array}{r}34\\4\overline{)136}\end{array}$	12) $\begin{array}{r}95\\8\overline{)760}\end{array}$

PAGE: 98

Division 5-Digit Dividends / 1-Digit Divisors

Divide.

1) $\begin{array}{r}21713\\4\overline{)86852}\end{array}$	2) $\begin{array}{r}9518\\3\overline{)28554}\end{array}$	3) $\begin{array}{r}12108\\4\overline{)48432}\end{array}$
4) $\begin{array}{r}10477\\3\overline{)31431}\end{array}$	5) $\begin{array}{r}4743\\9\overline{)42687}\end{array}$	6) $\begin{array}{r}9128\\9\overline{)82152}\end{array}$
7) $\begin{array}{r}14992\\2\overline{)29984}\end{array}$	8) $\begin{array}{r}6705\\2\overline{)13410}\end{array}$	9) $\begin{array}{r}20918\\2\overline{)41836}\end{array}$

PAGE: 99

Division 5-Digit Dividends / 1-Digit Divisors

Divide.

1) $\begin{array}{r}5135\\6\overline{)30810}\end{array}$	2) $\begin{array}{r}3974\\4\overline{)15896}\end{array}$	3) $\begin{array}{r}9956\\8\overline{)79648}\end{array}$
4) $\begin{array}{r}10235\\4\overline{)40940}\end{array}$	5) $\begin{array}{r}12088\\5\overline{)60440}\end{array}$	6) $\begin{array}{r}5329\\9\overline{)47961}\end{array}$
7) $\begin{array}{r}15280\\2\overline{)30560}\end{array}$	8) $\begin{array}{r}4884\\9\overline{)43956}\end{array}$	9) $\begin{array}{r}13610\\3\overline{)40830}\end{array}$

PAGE: 100

Division 5-Digit Dividends / 1-Digit Divisors

Divide.

1) 8258 2)16516	2) 4517 9)40653	3) 3396 5)16980
4) 13599 3)40797	5) 14145 2)28290	6) 3990 3)11970
7) 7548 2)15096	8) 18736 4)74944	9) 9495 9)85455

PAGE: 101

Division 5-Digit Dividends / 1-Digit Divisors

Divide.

1) 10936 7)76552	2) 17511 3)52533	3) 13222 7)92554
4) 18310 5)91550	5) 11732 6)70392	6) 9807 3)29421
7) 4800 5)24000	8) 20851 3)62553	9) 7792 6)46752

PAGE: 102

Division 5-Digit Dividends / 1-Digit Divisors

Divide.

1) 19884 2)39768	2) 12994 2)25988	3) 4085 7)28595
4) 4330 9)38970	5) 12180 7)85260	6) 16374 6)98244
7) 13645 3)40935	8) 19668 5)98340	9) 13462 3)40386

PAGE: 103

Division 5-Digit Dividends / 1-Digit Divisors

Divide.

1) 14561 6)87366	2) 15868 4)63472	3) 7441 5)37205
4) 17172 3)51516	5) 14433 4)57732	6) 10451 9)94059
7) 11620 3)34860	8) 15432 2)30864	9) 3479 3)10437

PAGE: 104

Division 5-Digit Dividends / 1-Digit Divisors

Divide.

1) 5812 9)52308	2) 12783 2)25566	3) 13605 6)81630
4) 15447 3)46341	5) 2489 9)22401	6) 14842 5)74210
7) 4516 3)13548	8) 15483 3)46449	9) 13899 3)41697

PAGE: 105

Division 5-Digit Dividends / 1-Digit Divisors

Divide.

1) 6437 9)57933	2) 7640 6)45840	3) 8160 3)24480
4) 18011 2)36022	5) 7354 9)66186	6) 19516 5)97580
7) 9726 7)68082	8) 9479 3)28437	9) 3433 8)27464

PAGE: 106

Division 5-Digit Dividends / 1-Digit Divisors

Divide.

1) 4)59324 = 14831
2) 3)31209 = 10403
3) 2)12522 = 6261
4) 4)76988 = 19247
5) 7)59563 = 8509
6) 8)97848 = 12231
7) 6)37296 = 6216
8) 6)22200 = 3700
9) 7)75516 = 10788

PAGE: 107

Division 5-Digit Dividends / 1-Digit Divisors

Divide.

1) 7)22645 = 3235
2) 4)69184 = 17296
3) 7)79884 = 11412
4) 7)35805 = 5115
5) 7)85071 = 12153
6) 4)80996 = 20249
7) 6)88296 = 14716
8) 4)17376 = 4344
9) 4)17952 = 4488

PAGE: 108

Division 5-Digit Dividends / 1-Digit Divisors

Divide.

1) 3)41601 = 13867
2) 4)82012 = 20503
3) 4)82888 = 20722
4) 4)72008 = 18002
5) 4)38584 = 9646
6) 6)14916 = 2486
7) 4)20468 = 5117
8) 6)86754 = 14459
9) 5)36895 = 7379

PAGE: 110

Division 4-Digit Dividends / 1-Digit Divisors *(remainders)*

Divide.

1) 7)3191 = 455R6
2) 5)3846 = 769R1
3) 6)7665 = 1277R3
4) 7)2774 = 396R2
5) 6)8221 = 1370R1
6) 2)9345 = 4672R1
7) 9)9038 = 1004R2
8) 6)7682 = 1280R2
9) 7)9838 = 1405R3

PAGE: 111

Division 4-Digit Dividends / 1-Digit Divisors *(remainders)*

Divide.

1) 9)4190 = 465R5
2) 2)5885 = 2942R1
3) 3)9734 = 3244R2
4) 8)5814 = 726R6
5) 8)3979 = 497R3
6) 9)2834 = 314R8
7) 6)6484 = 1080R4
8) 4)3826 = 956R2
9) 3)5272 = 1757R1

PAGE: 112

Division 4-Digit Dividends / 1-Digit Divisors *(remainders)*

Divide.

1) 6)7585 = 1264R1
2) 8)9476 = 1184R4
3) 2)9377 = 4688R1
4) 6)8924 = 1487R2
5) 3)1813 = 604R1
6) 5)1488 = 297R3
7) 6)4621 = 770R1
8) 8)8009 = 1001R1
9) 9)6136 = 681R7

PAGE: 113

Division 4-Digit Dividends / 1-Digit Divisors *(remainders)*

Divide.

1) 6)1363 = 227R1	2) 2)8007 = 4003R1	3) 5)7787 = 1557R2
4) 9)5836 = 648R4	5) 8)8492 = 1061R4	6) 6)8603 = 1433R5
7) 4)6550 = 1637R2	8) 8)6895 = 861R7	9) 9)7237 = 804R1

PAGE: 114

Division 4-Digit Dividends / 1-Digit Divisors *(remainders)*

Divide.

1) 7)2314 = 330R4	2) 2)2011 = 1005R1	3) 6)3340 = 556R4
4) 3)4456 = 1485R1	5) 5)7692 = 1538R2	6) 9)8416 = 935R1
7) 5)6926 = 1385R1	8) 8)2972 = 371R4	9) 4)6394 = 1598R2

PAGE: 115

Division 4-Digit Dividends / 1-Digit Divisors *(remainders)*

Divide.

1) 4)6714 = 1678R2	2) 7)8770 = 1252R6	3) 5)3753 = 750R3
4) 5)7182 = 1436R2	5) 3)7807 = 2602R1	6) 7)3813 = 544R5
7) 8)1662 = 207R6	8) 3)7711 = 2570R1	9) 8)3451 = 431R3

PAGE: 116

Division 4-Digit Dividends / 1-Digit Divisors *(remainders)*

Divide.

1) 3)6082 = 2027R1	2) 9)7907 = 878R5	3) 3)3182 = 1060R2
4) 6)3710 = 618R2	5) 5)5826 = 1165R1	6) 2)8357 = 4178R1
7) 5)1899 = 379R4	8) 8)4501 = 562R5	9) 9)6248 = 694R2

PAGE: 117

Division 4-Digit Dividends / 1-Digit Divisors *(remainders)*

Divide.

1) 7)6567 = 938R1	2) 6)8092 = 1348R4	3) 8)5892 = 736R4
4) 3)4123 = 1374R1	5) 6)2815 = 469R1	6) 3)8507 = 2835R2
7) 6)9867 = 1644R3	8) 8)3039 = 379R7	9) 4)7907 = 1976R3

PAGE: 118

Division 4-Digit Dividends / 1-Digit Divisors *(remainders)*

Divide.

1) 9)8418 = 935R3	2) 3)1573 = 524R1	3) 2)1645 = 822R1
4) 6)2164 = 360R4	5) 8)2958 = 369R6	6) 7)3573 = 510R3
7) 5)5522 = 1104R2	8) 2)2329 = 1164R1	9) 3)3275 = 1091R2

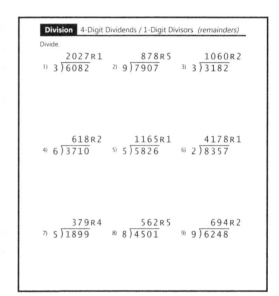

PAGE: 119

Division 4-Digit Dividends / 1-Digit Divisors *(remainders)*

Divide.

1) 7)5468 = 781R1	2) 5)8873 = 1774R3	3) 3)4667 = 1555R2
4) 9)8686 = 965R1	5) 9)2534 = 281R5	6) 8)4422 = 552R6
7) 7)1963 = 280R3	8) 5)5777 = 1155R2	9) 7)7115 = 1016R3

PAGE: 120

Division 4-Digit Dividends / 1-Digit Divisors *(remainders)*

Divide.

1) 5)5843 = 1168R3	2) 8)6782 = 847R6	3) 2)8725 = 4362R1
4) 9)5197 = 577R4	5) 9)9722 = 1080R2	6) 9)6709 = 745R4
7) 8)6084 = 760R4	8) 8)4793 = 599R1	9) 9)2662 = 295R7

PAGE: 122

Division 4-Digit Dividends / 2-Digit Divisors

Divide.

1) 51)4284 = 84	2) 29)5626 = 194	3) 54)9396 = 174
4) 10)9840 = 984	5) 46)5336 = 116	6) 15)5715 = 381
7) 21)5460 = 260	8) 71)6887 = 97	9) 41)2419 = 59

PAGE: 123

Division 4-Digit Dividends / 2-Digit Divisors

Divide.

1) 69)2898 = 42	2) 35)5075 = 145	3) 48)2208 = 46
4) 64)2432 = 38	5) 89)1691 = 19	6) 19)4579 = 241
7) 44)1100 = 25	8) 28)3416 = 122	9) 39)9906 = 254

PAGE: 124

Division 4-Digit Dividends / 2-Digit Divisors

Divide.

1) 27)4725 = 175	2) 21)3591 = 171	3) 16)5824 = 364
4) 80)1040 = 13	5) 40)4480 = 112	6) 85)5950 = 70
7) 20)5540 = 277	8) 33)8283 = 251	9) 31)4960 = 160

PAGE: 125

Division 4-Digit Dividends / 2-Digit Divisors

Divide.

1) 22)8888 = 404	2) 12)6480 = 540	3) 32)2272 = 71
4) 23)1518 = 66	5) 21)2100 = 100	6) 86)4558 = 53
7) 11)3388 = 308	8) 26)6864 = 264	9) 48)8208 = 171

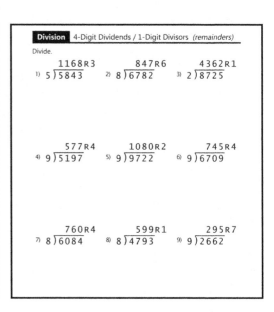

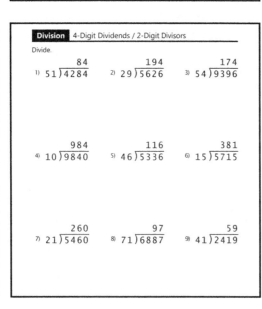

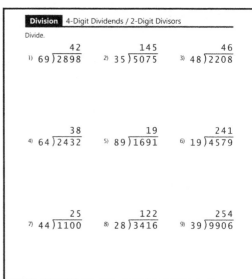

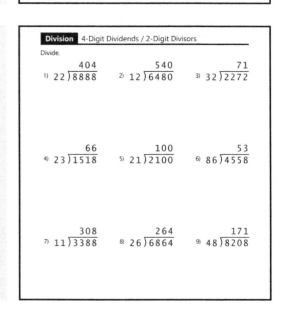

PAGE: 126

Division 4-Digit Dividends / 2-Digit Divisors

Divide.

1) 22)2618 → 119	2) 13)9581 → 737	3) 18)4014 → 223
4) 89)4895 → 55	5) 80)8160 → 102	6) 15)3855 → 257
7) 75)6075 → 81	8) 16)7248 → 453	9) 24)7152 → 298

PAGE: 127

Division 4-Digit Dividends / 2-Digit Divisors

Divide.

1) 88)9504 → 108	2) 11)7898 → 718	3) 70)7980 → 114
4) 10)7350 → 735	5) 12)7548 → 629	6) 45)7605 → 169
7) 28)8932 → 319	8) 64)4736 → 74	9) 57)2964 → 52

PAGE: 128

Division 4-Digit Dividends / 2-Digit Divisors

Divide.

1) 17)5287 → 311	2) 40)3080 → 77	3) 36)4608 → 128
4) 21)6216 → 296	5) 62)1178 → 19	6) 15)1440 → 96
7) 26)5928 → 228	8) 44)2684 → 61	9) 52)1612 → 31

PAGE: 129

Division 4-Digit Dividends / 2-Digit Divisors

Divide.

1) 31)6076 → 196	2) 77)4928 → 64	3) 21)8841 → 421
4) 56)7840 → 140	5) 37)8695 → 235	6) 61)4392 → 72
7) 55)6710 → 122	8) 40)9480 → 237	9) 45)5985 → 133

PAGE: 130

Division 4-Digit Dividends / 2-Digit Divisors

Divide.

1) 71)4331 → 61	2) 12)2628 → 219	3) 27)2322 → 86
4) 35)8190 → 234	5) 81)3969 → 49	6) 38)4218 → 111
7) 60)5340 → 89	8) 12)6048 → 504	9) 16)5248 → 328

PAGE: 131

Division 4-Digit Dividends / 2-Digit Divisors

Divide.

1) 31)8835 → 285	2) 20)1280 → 64	3) 34)6732 → 198
4) 11)2596 → 236	5) 10)3250 → 325	6) 60)7500 → 125
7) 87)7743 → 89	8) 79)5214 → 66	9) 33)7095 → 215

PAGE: 132

Division 4-Digit Dividends / 2-Digit Divisors

Divide.

1) 44)8184 = 186
2) 18)5292 = 294
3) 85)2890 = 34
4) 32)9728 = 304
5) 22)8096 = 368
6) 23)8073 = 351
7) 11)6523 = 593
8) 32)3424 = 107
9) 10)1640 = 164

PAGE: 134

Multiplication By 10, 100 and 1000

Multiply.

1) 59x100=	5900	11) 22x1000=	22000
2) 93x10=	930	12) 32x10=	320
3) 92x100=	9200	13) 69x10=	690
4) 42x10=	420	14) 14x1000=	14000
5) 58x10=	580	15) 44x1000=	44000
6) 41x10=	410	16) 40x10=	400
7) 20x1000=	20000	17) 95x10=	950
8) 68x1000=	68000	18) 90x100=	9000
9) 43x1000=	43000	19) 32x100=	3200
10) 18x1000=	18000	20) 57x100=	5700

PAGE: 135

Multiplication By 10, 100 and 1000

Multiply.

1) 21x10=	210	11) 87x10=	870
2) 38x10=	380	12) 89x1000=	89000
3) 16x100=	1600	13) 70x1000=	70000
4) 24x1000=	24000	14) 81x1000=	81000
5) 92x100=	9200	15) 45x100=	4500
6) 38x100=	3800	16) 96x10=	960
7) 60x10=	600	17) 74x1000=	74000
8) 59x100=	5900	18) 28x1000=	28000
9) 66x10=	660	19) 52x1000=	52000
10) 67x100=	6700	20) 26x100=	2600

PAGE: 136

Multiplication By 10, 100 and 1000

Multiply.

1) 90x1000=	90000	11) 43x1000=	43000
2) 39x100=	3900	12) 69x100=	6900
3) 26x10=	260	13) 26x100=	2600
4) 40x10=	400	14) 49x100=	4900
5) 23x100=	2300	15) 63x10=	630
6) 48x10=	480	16) 32x1000=	32000
7) 58x100=	5800	17) 49x10=	490
8) 42x10=	420	18) 80x100=	8000
9) 27x1000=	27000	19) 48x100=	4800
10) 86x1000=	86000	20) 88x10=	880

PAGE: 137

Multiplication By 10, 100 and 1000

Multiply.

1) 77x1000=	77000	11) 38x100=	3800
2) 73x100=	7300	12) 34x100=	3400
3) 18x100=	1800	13) 84x100=	8400
4) 68x100=	6800	14) 67x1000=	67000
5) 82x10=	820	15) 78x1000=	78000
6) 49x100=	4900	16) 81x10=	810
7) 80x1000=	80000	17) 85x10=	850
8) 79x10=	790	18) 55x100=	5500
9) 75x10=	750	19) 83x100=	8300
10) 53x1000=	53000	20) 51x1000=	51000

PAGE: 138

Multiplication By 10, 100 and 1000

Multiply.

1) 44x10=	440	11) 51x100=	5100
2) 35x100=	3500	12) 50x100=	5000
3) 53x100=	5300	13) 54x10=	540
4) 87x100=	8700	14) 29x100=	2900
5) 80x100=	8000	15) 69x1000=	69000
6) 20x100=	2000	16) 53x1000=	53000
7) 19x100=	1900	17) 75x10=	750
8) 65x1000=	65000	18) 15x1000=	15000
9) 18x1000=	18000	19) 34x1000=	34000
10) 64x10=	640	20) 56x100=	5600

PAGE: 139

Multiplication By 10, 100 and 1000

Multiply.

1) 81x100= 8100	11) 60x1000= 60000
2) 89x1000= 89000	12) 69x100= 6900
3) 16x1000= 16000	13) 38x100= 3800
4) 61x10= 610	14) 42x100= 4200
5) 18x10= 180	15) 73x1000= 73000
6) 87x10= 870	16) 55x100= 5500
7) 64x100= 6400	17) 17x10= 170
8) 41x100= 4100	18) 32x10= 320
9) 69x1000= 69000	19) 44x1000= 44000
10) 71x1000= 71000	20) 97x1000= 97000

PAGE: 140

Multiplication By 10, 100 and 1000

Multiply.

1) 45x100= 4500	11) 71x1000= 71000
2) 52x100= 5200	12) 43x100= 4300
3) 86x100= 8600	13) 20x10= 200
4) 51x10= 510	14) 54x100= 5400
5) 60x100= 6000	15) 67x100= 6700
6) 15x1000= 15000	16) 25x100= 2500
7) 31x100= 3100	17) 93x10= 930
8) 18x100= 1800	18) 62x1000= 62000
9) 28x10= 280	19) 93x1000= 93000
10) 14x1000= 14000	20) 47x10= 470

PAGE: 141

Multiplication By 10, 100 and 1000

Multiply.

1) 70x1000= 70000	11) 79x1000= 79000
2) 76x100= 7600	12) 50x1000= 50000
3) 40x100= 4000	13) 30x100= 3000
4) 97x10= 970	14) 50x10= 500
5) 86x10= 860	15) 95x1000= 95000
6) 57x1000= 57000	16) 25x1000= 25000
7) 17x100= 1700	17) 81x100= 8100
8) 18x1000= 18000	18) 39x1000= 39000
9) 67x10= 670	19) 54x10= 540
10) 45x10= 450	20) 71x1000= 71000

PAGE: 142

Multiplication By 10, 100 and 1000

Multiply.

1) 37x10= 370	11) 24x10= 240
2) 97x100= 9700	12) 67x100= 6700
3) 46x10= 460	13) 93x10= 930
4) 20x10= 200	14) 47x10= 470
5) 25x1000= 25000	15) 86x1000= 86000
6) 29x1000= 29000	16) 67x1000= 67000
7) 38x10= 380	17) 64x100= 6400
8) 19x100= 1900	18) 20x1000= 20000
9) 64x10= 640	19) 92x1000= 92000
10) 53x10= 530	20) 54x10= 540

PAGE: 143

Multiplication By 10, 100 and 1000

Multiply.

1) 80x100= 8000	11) 63x1000= 63000
2) 26x1000= 26000	12) 64x10= 640
3) 67x100= 6700	13) 42x100= 4200
4) 85x1000= 85000	14) 27x100= 2700
5) 88x1000= 88000	15) 92x1000= 92000
6) 41x100= 4100	16) 94x1000= 94000
7) 37x10= 370	17) 22x100= 2200
8) 41x1000= 41000	18) 73x1000= 73000
9) 61x100= 6100	19) 65x100= 6500
10) 56x1000= 56000	20) 30x1000= 30000

PAGE: 144

Multiplication By 10, 100 and 1000

Multiply.

1) 76x1000= 76000	11) 47x100= 4700
2) 82x10= 820	12) 53x100= 5300
3) 52x10= 520	13) 31x10= 310
4) 35x10= 350	14) 96x1000= 96000
5) 40x1000= 40000	15) 64x10= 640
6) 32x10= 320	16) 50x100= 5000
7) 19x100= 1900	17) 41x10= 410
8) 53x10= 530	18) 65x10= 650
9) 54x100= 5400	19) 25x100= 2500
10) 75x100= 7500	20) 34x100= 3400

PAGE: 146

Multiplication Find the Missing Multipliers (10, 100 or 1000)

Fill in the blanks with 10, 100 or 1000.

1) 11x 1000 =11000	11) 7x 10 =70
2) 69x 10 =690	12) 57x 1000 =57000
3) 97x 1000 =97000	13) 69x 100 =6900
4) 90x 10 =900	14) 5x 10 =50
5) 21x 100 =2100	15) 74x 10 =740
6) 2x 10 =20	16) 4x 10 =40
7) 32x 100 =3200	17) 60x 10 =600
8) 59x 10 =590	18) 79x 100 =7900
9) 3x 10 =30	19) 42x 10 =420
10) 78x 1000 =78000	20) 41x 100 =4100

PAGE: 147

Multiplication Find the Missing Multipliers (10, 100 or 1000)

Fill in the blanks with 10, 100 or 1000.

1) 73x 1000 =73000	11) 24x 100 =2400
2) 8x 10 =80	12) 86x 1000 =86000
3) 51x 10 =510	13) 79x 1000 =79000
4) 28x 1000 =28000	14) 95x 100 =9500
5) 10x 10 =100	15) 84x 100 =8400
6) 17x 100 =1700	16) 42x 10 =420
7) 23x 1000 =23000	17) 15x 1000 =15000
8) 48x 10 =480	18) 98x 1000 =98000
9) 65x 100 =6500	19) 43x 10 =430
10) 96x 10 =960	20) 43x 1000 =43000

PAGE: 148

Multiplication Find the Missing Multipliers (10, 100 or 1000)

Fill in the blanks with 10, 100 or 1000.

1) 16x 100 =1600	11) 46x 100 =4600
2) 40x 1000 =40000	12) 49x 1000 =49000
3) 45x 1000 =45000	13) 97x 100 =9700
4) 51x 100 =5100	14) 90x 10 =900
5) 30x 1000 =30000	15) 74x 1000 =74000
6) 5x 1000 =5000	16) 45x 100 =4500
7) 46x 1000 =46000	17) 50x 10 =500
8) 75x 1000 =75000	18) 85x 1000 =85000
9) 66x 1000 =66000	19) 98x 100 =9800
10) 40x 10 =400	20) 87x 1000 =87000

PAGE: 149

Multiplication Find the Missing Multipliers (10, 100 or 1000)

Fill in the blanks with 10, 100 or 1000.

1) 67x 1000 =67000	11) 97x 100 =9700
2) 72x 100 =7200	12) 28x 1000 =28000
3) 26x 10 =260	13) 23x 10 =230
4) 48x 1000 =48000	14) 78x 100 =7800
5) 14x 100 =1400	15) 45x 1000 =45000
6) 41x 1000 =41000	16) 17x 100 =1700
7) 7x 1000 =7000	17) 41x 10 =410
8) 17x 10 =170	18) 78x 10 =780
9) 86x 10 =860	19) 81x 10 =810
10) 52x 100 =5200	20) 9x 1000 =9000

PAGE: 150

Multiplication Find the Missing Multipliers (10, 100 or 1000)

Fill in the blanks with 10, 100 or 1000.

1) 57x 1000 =57000	11) 78x 10 =780
2) 98x 1000 =98000	12) 67x 100 =6700
3) 27x 100 =2700	13) 93x 10 =930
4) 35x 100 =3500	14) 32x 10 =320
5) 69x 10 =690	15) 4x 10 =40
6) 31x 100 =3100	16) 76x 1000 =76000
7) 62x 10 =620	17) 63x 1000 =63000
8) 77x 1000 =77000	18) 79x 1000 =79000
9) 96x 10 =960	19) 9x 100 =900
10) 10x 1000 =10000	20) 94x 100 =9400

PAGE: 151

Multiplication Find the Missing Multipliers (10, 100 or 1000)

Fill in the blanks with 10, 100 or 1000.

1) 35x 1000 =35000	11) 40x 1000 =40000
2) 46x 1000 =46000	12) 85x 10 =850
3) 64x 100 =6400	13) 33x 100 =3300
4) 63x 1000 =63000	14) 26x 100 =2600
5) 62x 100 =6200	15) 62x 10 =620
6) 36x 10 =360	16) 88x 100 =8800
7) 5x 100 =500	17) 39x 100 =3900
8) 53x 10 =530	18) 62x 1000 =62000
9) 47x 10 =470	19) 89x 1000 =89000
10) 16x 100 =1600	20) 23x 1000 =23000

PAGE: 152

Multiplication Find the Missing Multipliers (10, 100 or 1000)

Fill in the blanks with 10, 100 or 1000.

1) 78x 1000 =78000
2) 60x 10 =600
3) 80x 1000 =80000
4) 14x 100 =1400
5) 76x 100 =7600
6) 46x 10 =460
7) 41x 1000 =41000
8) 34x 100 =3400
9) 47x 10 =470
10) 52x 10 =520
11) 58x 10 =580
12) 30x 100 =3000
13) 80x 10 =800
14) 60x 100 =6000
15) 64x 100 =6400
16) 41x 10 =410
17) 84x 10 =840
18) 45x 1000 =45000
19) 18x 10 =180
20) 49x 10 =490

PAGE: 153

Multiplication Find the Missing Multipliers (10, 100 or 1000)

Fill in the blanks with 10, 100 or 1000.

1) 88x 1000 =88000
2) 6x 1000 =6000
3) 95x 1000 =95000
4) 37x 1000 =37000
5) 47x 10 =470
6) 50x 100 =5000
7) 10x 1000 =10000
8) 22x 1000 =22000
9) 83x 10 =830
10) 43x 1000 =43000
11) 56x 1000 =56000
12) 54x 10 =540
13) 70x 100 =7000
14) 52x 10 =520
15) 98x 10 =980
16) 16x 100 =1600
17) 67x 10 =670
18) 76x 100 =7600
19) 17x 10 =170
20) 7x 100 =700

PAGE: 154

Multiplication Find the Missing Multipliers (10, 100 or 1000)

Fill in the blanks with 10, 100 or 1000.

1) 93x 100 =9300
2) 86x 10 =860
3) 19x 100 =1900
4) 32x 100 =3200
5) 28x 1000 =28000
6) 50x 1000 =50000
7) 97x 1000 =97000
8) 19x 1000 =19000
9) 8x 100 =800
10) 11x 10 =110
11) 20x 100 =2000
12) 77x 100 =7700
13) 54x 10 =540
14) 68x 1000 =68000
15) 92x 100 =9200
16) 51x 1000 =51000
17) 15x 1000 =15000
18) 4x 100 =400
19) 88x 100 =8800
20) 95x 10 =950

PAGE: 155

Multiplication Find the Missing Multipliers (10, 100 or 1000)

Fill in the blanks with 10, 100 or 1000.

1) 27x 10 =270
2) 3x 100 =300
3) 33x 10 =330
4) 66x 10 =660
5) 2x 100 =200
6) 80x 10 =800
7) 7x 10 =70
8) 21x 100 =2100
9) 23x 10 =230
10) 58x 10 =580
11) 94x 100 =9400
12) 10x 100 =1000
13) 9x 10 =90
14) 15x 10 =150
15) 98x 1000 =98000
16) 86x 1000 =86000
17) 47x 1000 =47000
18) 4x 100 =400
19) 26x 10 =260
20) 37x 100 =3700

PAGE: 156

Multiplication Find the Missing Multipliers (10, 100 or 1000)

Fill in the blanks with 10, 100 or 1000.

1) 61x 10 =610
2) 71x 100 =7100
3) 60x 10 =600
4) 25x 1000 =25000
5) 11x 100 =1100
6) 15x 100 =1500
7) 82x 100 =8200
8) 16x 100 =1600
9) 55x 100 =5500
10) 4x 10 =40
11) 25x 100 =2500
12) 89x 100 =8900
13) 39x 100 =3900
14) 42x 1000 =42000
15) 22x 100 =2200
16) 48x 100 =4800
17) 46x 10 =460
18) 85x 100 =8500
19) 88x 100 =8800
20) 29x 100 =2900

PAGE: 158

Multiplication By 10, 100 and 1000 *(with decimals)*

Multiply.

1) .44x10= 4.4
2) 9.1x100= 910
3) 4.1x100= 410
4) .85x1000= 850
5) .057x10= .57
6) 4.5x1000= 4500
7) .022x100= 2.2
8) .71x100= 71
9) 1.2x100= 120
10) .53x1000= 530
11) 6.3x1000= 6300
12) .86x100= 86
13) .058x100= 5.8
14) .97x100= 97
15) 5.2x1000= 5200
16) .12x10= 1.2
17) .38x100= 38
18) .9x1000= 900
19) 5.3x100= 530
20) .54x10= 5.4

PAGE: 159

Multiplication By 10, 100 and 1000 *(with decimals)*

Multiply.

1) .9x1000=	900	11) .089x1000=	89
2) .064x10=	.64	12) 3.9x1000=	3900
3) 3.9x10=	39	13) 6.5x100=	650
4) 4.5x1000=	4500	14) .067x10=	.67
5) .043x10=	.43	15) .039x1000=	39
6) .09x1000=	90	16) .03x1000=	30
7) 9.1x100=	910	17) 7.7x10=	77
8) 2.3x100=	230	18) 2.9x100=	290
9) 5.2x100=	520	19) .61x1000=	610
10) .076x10=	.76	20) .18x1000=	180

PAGE: 160

Multiplication By 10, 100 and 1000 *(with decimals)*

Multiply.

1) .5x10=	5	11) .093x100=	9.3
2) .5x1000=	500	12) .47x1000=	470
3) .14x10=	1.4	13) .17x100=	17
4) .057x100=	5.7	14) 6.7x1000=	6700
5) .33x1000=	330	15) .041x1000=	41
6) .094x10=	.94	16) .68x100=	68
7) .01x100=	1	17) .018x100=	1.8
8) .64x1000=	640	18) .63x10=	6.3
9) .13x100=	13	19) .26x10=	2.6
10) 2.3x100=	230	20) .081x100=	8.1

PAGE: 161

Multiplication By 10, 100 and 1000 *(with decimals)*

Multiply.

1) .88x10=	8.8	11) .072x10=	.72
2) .055x100=	5.5	12) 2.3x10=	23
3) .032x10=	.32	13) .063x1000=	63
4) 3.1x1000=	3100	14) .63x100=	63
5) .5x10=	5	15) 6.2x100=	620
6) .34x10=	3.4	16) 2.7x100=	270
7) 3.6x10=	36	17) 1.2x10=	12
8) .01x1000=	10	18) .73x10=	7.3
9) 1.5x10=	15	19) 2.3x100=	230
10) .021x1000=	21	20) .066x10=	.66

PAGE: 162

Multiplication By 10, 100 and 1000 *(with decimals)*

Multiply.

1) .014x10=	.14	11) 4.7x1000=	4700
2) .098x100=	9.8	12) 6.7x100=	670
3) .33x10=	3.3	13) 1.4x100=	140
4) .038x1000=	38	14) .079x100=	7.9
5) 3.8x100=	380	15) 9.7x1000=	9700
6) 7.6x1000=	7600	16) 5.7x1000=	5700
7) .045x100=	4.5	17) .032x10=	.32
8) .06x10=	.6	18) .4x10=	4
9) .019x100=	1.9	19) .89x10=	8.9
10) .064x1000=	64	20) 6.8x10=	68

PAGE: 163

Multiplication By 10, 100 and 1000 *(with decimals)*

Multiply.

1) .74x100=	74	11) .3x1000=	300
2) 4.7x1000=	4700	12) .059x10=	.59
3) .78x1000=	780	13) .46x10=	4.6
4) 8.9x1000=	8900	14) .027x100=	2.7
5) .47x10=	4.7	15) .05x1000=	50
6) 8.6x10=	86	16) .73x100=	73
7) .45x100=	45	17) .091x10=	.91
8) 7.6x1000=	7600	18) .55x100=	55
9) .087x1000=	87	19) .98x100=	98
10) .52x10=	5.2	20) .72x100=	72

PAGE: 164

Multiplication By 10, 100 and 1000 *(with decimals)*

Multiply.

1) .042x10=	.42	11) .52x10=	5.2
2) .51x1000=	510	12) .079x10=	.79
3) .22x1000=	220	13) 7.7x1000=	7700
4) .65x1000=	650	14) .79x10=	7.9
5) .092x10=	.92	15) 5.3x1000=	5300
6) .037x10=	.37	16) .026x10=	.26
7) 7.2x1000=	7200	17) .09x100=	9
8) .54x100=	54	18) 2.1x10=	21
9) 2.7x10=	27	19) .2x1000=	200
10) 5.8x1000=	5800	20) 1.8x1000=	1800

PAGE: 165

Multiplication By 10, 100 and 1000 *(with decimals)*

Multiply.

1) 6.7x10= 67	11) 1.6x10= 16
2) .042x10= .42	12) .064x1000= 64
3) .73x10= 7.3	13) .043x100= 4.3
4) .79x100= 79	14) .031x1000= 31
5) .025x100= 2.5	15) .53x10= 5.3
6) .04x100= 4	16) .33x1000= 330
7) .044x100= 4.4	17) 1.4x1000= 1400
8) .039x100= 3.9	18) .85x1000= 850
9) .61x100= 61	19) .015x1000= 15
10) .92x10= 9.2	20) 5.6x10= 56

PAGE: 166

Multiplication By 10, 100 and 1000 *(with decimals)*

Multiply.

1) .054x10= .54	11) 9.4x10= 94
2) .2x1000= 200	12) 5.9x100= 590
3) .05x1000= 50	13) .49x100= 49
4) .2x10= 2	14) .061x100= 6.1
5) .029x100= 2.9	15) .16x10= 1.6
6) .46x10= 4.6	16) 7.5x1000= 7500
7) 1.8x10= 18	17) .082x10= .82
8) .04x10= .4	18) .7x100= 70
9) 9.2x1000= 9200	19) .094x1000= 94
10) .43x10= 4.3	20) 8.5x100= 850

PAGE: 167

Multiplication By 10, 100 and 1000 *(with decimals)*

Multiply.

1) .042x100= 4.2	11) 6.9x100= 690
2) .12x10= 1.2	12) .072x100= 7.2
3) 6.3x1000= 6300	13) .36x1000= 360
4) .038x10= .38	14) .81x10= 8.1
5) .05x10= .5	15) .45x1000= 450
6) .008x1000= 8	16) .91x100= 91
7) 4.2x100= 420	17) .051x100= 5.1
8) .34x100= 34	18) .032x10= .32
9) .32x1000= 320	19) .55x10= 5.5
10) 8.2x100= 820	20) .076x10= .76

PAGE: 168

Multiplication By 10, 100 and 1000 *(with decimals)*

Multiply.

1) 5.8x10= 58	11) .12x100= 12
2) .071x1000= 71	12) 5.7x100= 570
3) .14x1000= 140	13) .024x10= .24
4) 6.5x1000= 6500	14) .051x1000= 51
5) .36x10= 3.6	15) 4.3x1000= 4300
6) .077x10= .77	16) 8.4x10= 84
7) .019x10= .19	17) .074x10= .74
8) 7.6x100= 760	18) .4x100= 40
9) .02x1000= 20	19) .77x100= 77
10) .41x1000= 410	20) .072x1000= 72

PAGE: 170

Multiplication Find the Missing Multipliers (10, 100 or 1000)

Fill in the blanks with 10, 100 or 1000.

1) 6.9x 100 =690	11) 5.4x 100 =540
2) .19x 1000 =190	12) .13x 100 =13
3) .79x 1000 =790	13) .091x 10 =.91
4) .74x 1000 =740	14) .07x 100 =7
5) .097x 10 =.97	15) .058x 100 =5.8
6) .02x 1000 =20	16) 8.4x 100 =840
7) 6.3x 10 =63	17) .032x 10 =.32
8) .42x 10 =4.2	18) 4.7x 100 =470
9) .76x 10 =7.6	19) .77x 100 =77
10) 2.3x 1000 =2300	20) .094x 100 =9.4

PAGE: 171

Multiplication Find the Missing Multipliers (10, 100 or 1000)

Fill in the blanks with 10, 100 or 1000.

1) .91x 10 =9.1	11) 1.5x 10 =15
2) 8.6x 1000 =8600	12) .022x 100 =2.2
3) 1.5x 1000 =1500	13) .005x 1000 =5
4) 1.6x 1000 =1600	14) .066x 100 =6.6
5) .92x 100 =92	15) .63x 10 =6.3
6) .035x 10 =.35	16) .021x 1000 =21
7) .34x 10 =3.4	17) .059x 10 =.59
8) .023x 10 =.23	18) .25x 100 =25
9) .088x 100 =8.8	19) .08x 100 =8
10) .68x 1000 =680	20) .62x 100 =62

PAGE: 172

Multiplication Find the Missing Multipliers (10, 100 or 1000)

Fill in the blanks with 10, 100 or 1000.

1) .07x 10 =.7	11) 6.9x 100 =690
2) 9.3x 1000 =9300	12) .8x 100 =80
3) .67x 10 =6.7	13) .36x 10 =3.6
4) .9x 1000 =900	14) 2.6x 100 =260
5) .062x 100 =6.2	15) .081x 1000 =81
6) .02x 100 =2	16) .071x 10 =.71
7) .04x 100 =4	17) 1.4x 10 =14
8) 4.7x 100 =470	18) .06x 10 =.6
9) 8.3x 10 =83	19) .69x 10 =6.9
10) .007x 1000 =7	20) 4.4x 1000 =4400

PAGE: 173

Multiplication Find the Missing Multipliers (10, 100 or 1000)

Fill in the blanks with 10, 100 or 1000.

1) .46x 100 =46	11) .059x 100 =5.9
2) .51x 100 =51	12) 8.7x 1000 =8700
3) .03x 100 =3	13) .025x 100 =2.5
4) .037x 10 =.37	14) .079x 10 =.79
5) .054x 1000 =54	15) .064x 1000 =64
6) .071x 10 =.71	16) 5.8x 100 =580
7) 8.7x 1000 =8700	17) .9x 1000 =900
8) .065x 10 =.65	18) .025x 1000 =25
9) .013x 1000 =13	19) .036x 10 =.36
10) 7.9x 10 =79	20) .093x 1000 =93

PAGE: 174

Multiplication Find the Missing Multipliers (10, 100 or 1000)

Fill in the blanks with 10, 100 or 1000.

1) .085x 100 =8.5	11) .019x 100 =1.9
2) 9.1x 10 =91	12) 1.8x 100 =180
3) .6x 1000 =600	13) .59x 10 =5.9
4) .55x 1000 =550	14) 1.3x 10 =13
5) 9.4x 1000 =9400	15) 8.5x 1000 =8500
6) .007x 100 =.7	16) .06x 1000 =60
7) .019x 10 =.19	17) .013x 10 =.13
8) .022x 1000 =22	18) .97x 10 =9.7
9) 8.7x 10 =87	19) .28x 1000 =280
10) .5x 1000 =500	20) .59x 100 =59

PAGE: 175

Multiplication Find the Missing Multipliers (10, 100 or 1000)

Fill in the blanks with 10, 100 or 1000.

1) .98x 10 =9.8	11) .097x 1000 =97
2) .009x 100 =.9	12) 7.1x 100 =710
3) 8.5x 10 =85	13) .043x 10 =.43
4) .075x 1000 =75	14) .035x 10 =.35
5) .87x 100 =87	15) 7.9x 10 =79
6) .35x 10 =3.5	16) .048x 1000 =48
7) 4.7x 10 =47	17) .63x 1000 =630
8) .4x 10 =4	18) .04x 10 =.4
9) 5.2x 1000 =5200	19) .095x 1000 =95
10) .43x 1000 =430	20) .087x 1000 =87

PAGE: 176

Multiplication Find the Missing Multipliers (10, 100 or 1000)

Fill in the blanks with 10, 100 or 1000.

1) 4.4x 100 =440	11) .031x 100 =3.1
2) 7.1x 100 =710	12) 2.3x 1000 =2300
3) 1.5x 10 =15	13) .7x 1000 =700
4) 5.5x 100 =550	14) .19x 100 =19
5) .04x 10 =.4	15) .68x 100 =68
6) .11x 1000 =110	16) .008x 1000 =8
7) .002x 10 =.02	17) .29x 100 =29
8) .97x 10 =9.7	18) .63x 1000 =630
9) .48x 10 =4.8	19) .004x 100 =.4
10) .04x 100 =4	20) .019x 1000 =19

PAGE: 177

Multiplication Find the Missing Multipliers (10, 100 or 1000)

Fill in the blanks with 10, 100 or 1000.

1) .037x 100 =3.7	11) 9.3x 1000 =9300
2) .096x 100 =9.6	12) .49x 10 =4.9
3) .07x 10 =.7	13) 5.7x 100 =570
4) .87x 1000 =870	14) 8.2x 1000 =8200
5) .047x 1000 =47	15) .77x 100 =77
6) .097x 100 =9.7	16) .98x 10 =9.8
7) .058x 1000 =58	17) 7.9x 10 =79
8) .25x 1000 =250	18) 5.6x 100 =560
9) .053x 100 =5.3	19) .062x 10 =.62
10) 3.8x 100 =380	20) .055x 10 =.55

PAGE: 178

Multiplication Find the Missing Multipliers (10, 100 or 1000)

Fill in the blanks with 10, 100 or 1000.

1) .32x 100 =32	11) .04x 10 =.4
2) .068x 1000 =68	12) 8.8x 100 =880
3) 9.4x 100 =940	13) .77x 100 =77
4) .91x 1000 =910	14) .97x 1000 =970
5) .98x 1000 =980	15) 9.1x 1000 =9100
6) .71x 1000 =710	16) 4.6x 1000 =4600
7) .48x 1000 =480	17) .086x 100 =8.6
8) 5.2x 100 =520	18) .88x 10 =8.8
9) .062x 1000 =62	19) 6.3x 10 =63
10) 3.1x 100 =310	20) .098x 10 =.98

PAGE: 179

Multiplication Find the Missing Multipliers (10, 100 or 1000)

Fill in the blanks with 10, 100 or 1000.

1) .046x 1000 =46	11) 4.3x 1000 =4300
2) 2.5x 1000 =2500	12) .075x 1000 =75
3) .64x 10 =6.4	13) .084x 1000 =84
4) 7.3x 100 =730	14) .56x 100 =56
5) .94x 10 =9.4	15) 9.5x 1000 =9500
6) .32x 10 =3.2	16) .57x 10 =5.7
7) .067x 100 =6.7	17) .065x 10 =.65
8) .002x 10 =.02	18) 4.4x 1000 =4400
9) .59x 1000 =590	19) .036x 10 =.36
10) .034x 1000 =34	20) .032x 100 =3.2

PAGE: 180

Multiplication Find the Missing Multipliers (10, 100 or 1000)

Fill in the blanks with 10, 100 or 1000.

1) 9.7x 1000 =9700	11) .094x 100 =9.4
2) 3.6x 100 =360	12) .085x 1000 =85
3) 6.2x 1000 =6200	13) .85x 10 =8.5
4) .057x 1000 =57	14) .25x 100 =25
5) .98x 10 =9.8	15) .5x 10 =5
6) .11x 100 =11	16) .032x 100 =3.2
7) 7.3x 10 =73	17) 2.8x 1000 =2800
8) .59x 10 =5.9	18) .42x 1000 =420
9) .04x 1000 =40	19) .32x 1000 =320
10) .039x 10 =.39	20) .089x 1000 =89

PAGE: 182

Division Divide by 10, 100 and 1000

Divide.

1) 3200÷100= 32	11) 2000÷1000= 2
2) 740÷10= 74	12) 12000÷1000= 12
3) 3800÷100= 38	13) 330÷10= 33
4) 9200÷100= 92	14) 1500÷100= 15
5) 62000÷1000= 62	15) 52000÷1000= 52
6) 180÷10= 18	16) 91000÷1000= 91
7) 79000÷1000= 79	17) 230÷10= 23
8) 200÷10= 20	18) 6500÷100= 65
9) 5000÷100= 50	19) 810÷10= 81
10) 70000÷1000= 70	20) 9300÷100= 93

PAGE: 183

Division Divide by 10, 100 and 1000

Divide.

1) 78000÷1000= 78	11) 72000÷1000= 72
2) 2200÷100= 22	12) 41000÷1000= 41
3) 9300÷100= 93	13) 680÷10= 68
4) 4600÷100= 46	14) 79000÷1000= 79
5) 7100÷100= 71	15) 45000÷1000= 45
6) 210÷10= 21	16) 8600÷100= 86
7) 85000÷1000= 85	17) 7700÷100= 77
8) 190÷10= 19	18) 38000÷1000= 38
9) 13000÷1000= 13	19) 40000÷1000= 40
10) 170÷10= 17	20) 2600÷100= 26

PAGE: 184

Division Divide by 10, 100 and 1000

Divide.

1) 82000÷1000= 82	11) 6800÷100= 68
2) 4300÷100= 43	12) 22000÷1000= 22
3) 390÷10= 39	13) 280÷10= 28
4) 910÷10= 91	14) 820÷10= 82
5) 74000÷1000= 74	15) 3600÷100= 36
6) 1000÷100= 10	16) 8100÷100= 81
7) 650÷10= 65	17) 680÷10= 68
8) 91000÷1000= 91	18) 87000÷1000= 87
9) 46000÷1000= 46	19) 8800÷100= 88
10) 1600÷100= 16	20) 7000÷1000= 7

PAGE: 185

Division Divide by 10, 100 and 1000

Divide.

1) 950÷10= 95	11) 900÷10= 90
2) 810÷10= 81	12) 90000÷1000= 90
3) 390÷10= 39	13) 18000÷1000= 18
4) 64000÷1000= 64	14) 4200÷100= 42
5) 700÷100= 7	15) 86000÷1000= 86
6) 590÷10= 59	16) 24000÷1000= 24
7) 720÷10= 72	17) 730÷10= 73
8) 410÷10= 41	18) 1100÷100= 11
9) 14000÷1000= 14	19) 540÷10= 54
10) 8300÷100= 83	20) 76000÷1000= 76

PAGE: 186

Division Divide by 10, 100 and 1000

Divide.

1) 640÷10= 64	11) 98000÷1000= 98
2) 17000÷1000= 17	12) 910÷10= 91
3) 5500÷100= 55	13) 53000÷1000= 53
4) 30000÷1000= 30	14) 3000÷1000= 3
5) 340÷10= 34	15) 130÷10= 13
6) 18000÷1000= 18	16) 5000÷1000= 5
7) 7300÷100= 73	17) 500÷10= 50
8) 5300÷100= 53	18) 73000÷1000= 73
9) 51000÷1000= 51	19) 45000÷1000= 45
10) 15000÷1000= 15	20) 850÷10= 85

PAGE: 187

Division Divide by 10, 100 and 1000

Divide.

1) 8100÷100= 81	11) 76000÷1000= 76
2) 860÷10= 86	12) 6500÷100= 65
3) 1700÷100= 17	13) 59000÷1000= 59
4) 1800÷100= 18	14) 7900÷100= 79
5) 8500÷100= 85	15) 96000÷1000= 96
6) 7100÷100= 71	16) 37000÷1000= 37
7) 680÷10= 68	17) 6600÷100= 66
8) 2600÷100= 26	18) 1200÷100= 12
9) 38000÷1000= 38	19) 41000÷1000= 41
10) 5300÷100= 53	20) 780÷10= 78

PAGE: 188

Division Divide by 10, 100 and 1000

Divide.

1) 6700÷100= 67	11) 160÷10= 16
2) 830÷10= 83	12) 95000÷1000= 95
3) 7200÷100= 72	13) 5000÷1000= 5
4) 30000÷1000= 30	14) 1100÷100= 11
5) 360÷10= 36	15) 8500÷100= 85
6) 530÷10= 53	16) 31000÷1000= 31
7) 8400÷100= 84	17) 3100÷100= 31
8) 34000÷1000= 34	18) 28000÷1000= 28
9) 370÷10= 37	19) 3300÷100= 33
10) 840÷10= 84	20) 38000÷1000= 38

PAGE: 189

Division Divide by 10, 100 and 1000

Divide.

1) 1900÷100= 19	11) 64000÷1000= 64
2) 500÷10= 50	12) 6500÷100= 65
3) 5100÷100= 51	13) 650÷10= 65
4) 670÷10= 67	14) 840÷10= 84
5) 9400÷100= 94	15) 95000÷1000= 95
6) 820÷10= 82	16) 230÷10= 23
7) 280÷10= 28	17) 340÷10= 34
8) 7300÷100= 73	18) 52000÷1000= 52
9) 21000÷1000= 21	19) 50000÷1000= 50
10) 140÷10= 14	20) 2400÷100= 24

PAGE: 190

Division Divide by 10, 100 and 1000

Divide.

1) 7600÷100= 76	11) 90000÷1000= 90
2) 270÷10= 27	12) 40000÷1000= 40
3) 5000÷100= 50	13) 210÷10= 21
4) 35000÷1000= 35	14) 7700÷100= 77
5) 400÷100= 4	15) 11000÷1000= 11
6) 250÷10= 25	16) 6400÷100= 64
7) 23000÷1000= 23	17) 3700÷100= 37
8) 6800÷100= 68	18) 740÷10= 74
9) 8400÷100= 84	19) 4400÷100= 44
10) 840÷10= 84	20) 6500÷100= 65

PAGE: 191

Division Divide by 10, 100 and 1000

Divide.

1) 82000÷1000= 82
2) 1600÷100= 16
3) 62000÷1000= 62
4) 23000÷1000= 23
5) 420÷10= 42
6) 27000÷1000= 27
7) 7100÷100= 71
8) 950÷10= 95
9) 6400÷100= 64
10) 2400÷100= 24
11) 52000÷1000= 52
12) 340÷10= 34
13) 3400÷100= 34
14) 270÷10= 27
15) 1700÷100= 17
16) 930÷10= 93
17) 700÷100= 7
18) 85000÷1000= 85
19) 4100÷100= 41
20) 8900÷100= 89

PAGE: 192

Division Divide by 10, 100 and 1000

Divide.

1) 610÷10= 61
2) 3700÷100= 37
3) 2900÷100= 29
4) 32000÷1000= 32
5) 930÷10= 93
6) 14000÷1000= 14
7) 35000÷1000= 35
8) 560÷10= 56
9) 97000÷1000= 97
10) 470÷10= 47
11) 1500÷100= 15
12) 88000÷1000= 88
13) 82000÷1000= 82
14) 690÷10= 69
15) 70000÷1000= 70
16) 57000÷1000= 57
17) 510÷10= 51
18) 250÷10= 25
19) 36000÷1000= 36
20) 380÷10= 38

PAGE: 194

Division Find the Missing Divisors (10, 100 or 1000)

Fill in the blanks with 10, 100 or 1000.

1) 38000÷ 1000 =38
2) 490÷ 10 =49
3) 3300÷ 100 =33
4) 97000÷ 1000 =97
5) 600÷ 100 =6
6) 6200÷ 100 =62
7) 300÷ 10 =30
8) 17000÷ 1000 =17
9) 8100÷ 100 =81
10) 3600÷ 100 =36
11) 3800÷ 100 =38
12) 740÷ 10 =74
13) 21000÷ 1000 =21
14) 58000÷ 1000 =58
15) 2500÷ 100 =25
16) 200÷ 100 =2
17) 240÷ 10 =24
18) 760÷ 10 =76
19) 93000÷ 1000 =93
20) 6100÷ 100 =61

PAGE: 195

Division Find the Missing Divisors (10, 100 or 1000)

Fill in the blanks with 10, 100 or 1000.

1) 6300÷ 100 =63
2) 4500÷ 100 =45
3) 750÷ 10 =75
4) 63000÷ 1000 =63
5) 940÷ 10 =94
6) 72000÷ 1000 =72
7) 670÷ 10 =67
8) 320÷ 10 =32
9) 2200÷ 100 =22
10) 2900÷ 100 =29
11) 680÷ 10 =68
12) 95000÷ 1000 =95
13) 540÷ 10 =54
14) 250÷ 10 =25
15) 8400÷ 100 =84
16) 83000÷ 1000 =83
17) 67000÷ 1000 =67
18) 240÷ 10 =24
19) 40000÷ 1000 =40
20) 85000÷ 1000 =85

PAGE: 196

Division Find the Missing Divisors (10, 100 or 1000)

Fill in the blanks with 10, 100 or 1000.

1) 17000÷ 1000 =17
2) 4200÷ 100 =42
3) 5900÷ 100 =59
4) 90000÷ 1000 =90
5) 1000÷ 100 =10
6) 340÷ 10 =34
7) 39000÷ 1000 =39
8) 450÷ 10 =45
9) 24000÷ 1000 =24
10) 9600÷ 100 =96
11) 580÷ 10 =58
12) 430÷ 10 =43
13) 86000÷ 1000 =86
14) 8600÷ 100 =86
15) 8800÷ 100 =88
16) 91000÷ 1000 =91
17) 42000÷ 1000 =42
18) 3500÷ 100 =35
19) 19000÷ 1000 =19
20) 88000÷ 1000 =88

PAGE: 197

Division Find the Missing Divisors (10, 100 or 1000)

Fill in the blanks with 10, 100 or 1000.

1) 390÷ 10 =39
2) 69000÷ 1000 =69
3) 6000÷ 100 =60
4) 340÷ 10 =34
5) 6100÷ 100 =61
6) 6600÷ 100 =66
7) 2000÷ 100 =20
8) 42000÷ 1000 =42
9) 87000÷ 1000 =87
10) 60000÷ 1000 =60
11) 920÷ 10 =92
12) 30000÷ 1000 =30
13) 440÷ 10 =44
14) 3300÷ 100 =33
15) 43000÷ 1000 =43
16) 71000÷ 1000 =71
17) 650÷ 10 =65
18) 230÷ 10 =23
19) 3500÷ 100 =35
20) 980÷ 10 =98

PAGE: 198

Division Find the Missing Divisors (10, 100 or 1000)

Fill in the blanks with 10, 100 or 1000.

1) 5600÷ 100 =56
2) 26000÷ 1000 =26
3) 44000÷ 1000 =44
4) 580÷ 10 =58
5) 6600÷ 10 =660
6) 8200÷ 100 =82
7) 7200÷ 10 =720
8) 250÷ 10 =25
9) 5000÷ 100 =50
10) 900÷ 100 =9
11) 3100÷ 10 =310
12) 46000÷ 1000 =46
13) 900÷ 10 =90
14) 860÷ 10 =86
15) 170÷ 10 =17
16) 54000÷ 1000 =54
17) 740÷ 10 =74
18) 8000÷ 10 =800
19) 700÷ 10 =70
20) 400÷ 10 =40

PAGE: 199

Division Find the Missing Divisors (10, 100 or 1000)

Fill in the blanks with 10, 100 or 1000.

1) 8300÷ 100 =83
2) 290÷ 10 =29
3) 87000÷ 1000 =87
4) 6000÷ 100 =60
5) 910÷ 10 =91
6) 7900÷ 100 =79
7) 82000÷ 1000 =82
8) 960÷ 10 =96
9) 77000÷ 1000 =77
10) 49000÷ 1000 =49
11) 35000÷ 1000 =35
12) 28000÷ 1000 =28
13) 41000÷ 1000 =41
14) 460÷ 10 =46
15) 65000÷ 1000 =65
16) 70000÷ 1000 =70
17) 74000÷ 1000 =74
18) 6800÷ 100 =68
19) 36000÷ 1000 =36
20) 9600÷ 100 =96

PAGE: 200

Division Find the Missing Divisors (10, 100 or 1000)

Fill in the blanks with 10, 100 or 1000.

1) 2700÷ 100 =27
2) 44000÷ 1000 =44
3) 2500÷ 100 =25
4) 2000÷ 1000 =2
5) 97000÷ 1000 =97
6) 150÷ 10 =15
7) 95000÷ 1000 =95
8) 410÷ 10 =41
9) 920÷ 10 =92
10) 6100÷ 100 =61
11) 9800÷ 100 =98
12) 23000÷ 1000 =23
13) 820÷ 10 =82
14) 9700÷ 100 =97
15) 89000÷ 1000 =89
16) 310÷ 10 =31
17) 41000÷ 1000 =41
18) 9200÷ 100 =92
19) 53000÷ 1000 =53
20) 6500÷ 100 =65

PAGE: 201

Division Find the Missing Divisors (10, 100 or 1000)

Fill in the blanks with 10, 100 or 1000.

1) 20000÷ 1000 =20
2) 68000÷ 1000 =68
3) 1500÷ 100 =15
4) 350÷ 10 =35
5) 7500÷ 100 =75
6) 36000÷ 1000 =36
7) 61000÷ 1000 =61
8) 490÷ 10 =49
9) 4600÷ 100 =46
10) 90000÷ 1000 =90
11) 31000÷ 1000 =31
12) 400÷ 100 =4
13) 5700÷ 100 =57
14) 6600÷ 100 =66
15) 5200÷ 100 =52
16) 8500÷ 100 =85
17) 890÷ 10 =89
18) 3700÷ 100 =37
19) 46000÷ 1000 =46
20) 9400÷ 100 =94

PAGE: 202

Division Find the Missing Divisors (10, 100 or 1000)

Fill in the blanks with 10, 100 or 1000.

1) 7100÷ 100 =71
2) 2000÷ 100 =20
3) 7800÷ 100 =78
4) 96000÷ 1000 =96
5) 330÷ 10 =33
6) 290÷ 10 =29
7) 530÷ 10 =53
8) 2200÷ 100 =22
9) 95000÷ 1000 =95
10) 140÷ 10 =14
11) 4000÷ 100 =40
12) 40000÷ 1000 =40
13) 3300÷ 100 =33
14) 720÷ 10 =72
15) 510÷ 10 =51
16) 13000÷ 1000 =13
17) 910÷ 10 =91
18) 4700÷ 100 =47
19) 260÷ 10 =26
20) 19000÷ 1000 =19

PAGE: 203

Division Find the Missing Divisors (10, 100 or 1000)

Fill in the blanks with 10, 100 or 1000.

1) 920÷ 10 =92
2) 8300÷ 100 =83
3) 8700÷ 100 =87
4) 500÷ 10 =50
5) 69000÷ 1000 =69
6) 90000÷ 1000 =90
7) 130÷ 10 =13
8) 800÷ 10 =80
9) 5300÷ 100 =53
10) 9500÷ 100 =95
11) 940÷ 10 =94
12) 89000÷ 1000 =89
13) 7600÷ 100 =76
14) 760÷ 10 =76
15) 630÷ 10 =63
16) 5700÷ 100 =57
17) 3100÷ 100 =31
18) 9100÷ 100 =91
19) 93000÷ 1000 =93
20) 83000÷ 1000 =83

PAGE: 204

Division Find the Missing Divisors (10, 100 or 1000)

Fill in the blanks with 10, 100 or 1000.

1) 9400÷ 100 =94
2) 16000÷ 1000 =16
3) 490÷ 10 =49
4) 95000÷ 1000 =95
5) 500÷ 100 =5
6) 650÷ 10 =65
7) 3200÷ 100 =32
8) 49000÷ 1000 =49
9) 63000÷ 1000 =63
10) 940÷ 10 =94
11) 8200÷ 100 =82
12) 5100÷ 100 =51
13) 400÷ 10 =40
14) 210÷ 10 =21
15) 15000÷ 1000 =15
16) 9300÷ 100 =93
17) 810÷ 10 =81
18) 1800÷ 100 =18
19) 5800÷ 100 =58
20) 4300÷ 100 =43

PAGE: 206

Division Divide by 10, 100 and 1000 *(with decimals)*

Divide.

1) 9.2÷100= .092
2) .75÷10= .075
3) 3.5÷10= .35
4) 64÷1000= .064
5) 8.6÷100= .086
6) 67÷100= .67
7) 44÷1000= .044
8) 750÷100= 7.5
9) 6.8÷10= .68
10) .7÷100= .007
11) 9.4÷100= .094
12) 2.1÷100= .021
13) 33÷1000= .033
14) .36÷10= .036
15) 9.5÷100= .095
16) 6100÷1000= 6.1
17) 8.9÷100= .089
18) 430÷100= 4.3
19) 280÷100= 2.8
20) 760÷1000= .76

PAGE: 207

Division Divide by 10, 100 and 1000 *(with decimals)*

Divide.

1) 20÷100= .2
2) 94÷100= .94
3) 8.6÷100= .086
4) 96÷1000= .096
5) 61÷100= .61
6) 470÷100= 4.7
7) .89÷10= .089
8) 260÷100= 2.6
9) 120÷1000= .12
10) 5÷100= .05
11) 7.3÷100= .073
12) 13÷1000= .013
13) 42÷10= 4.2
14) 6.8÷10= .68
15) 65÷1000= .065
16) 7800÷1000= 7.8
17) 680÷1000= .68
18) 28÷1000= .028
19) 70÷100= .7
20) .42÷10= .042

PAGE: 208

Division Divide by 10, 100 and 1000 *(with decimals)*

Divide.

1) 18÷100= .18
2) 8.4÷10= .84
3) 48÷1000= .048
4) 460÷1000= .46
5) 9.2÷10= .92
6) 98÷100= .98
7) 94÷100= .94
8) .74÷10= .074
9) 53÷1000= .053
10) 90÷100= .9
11) 7900÷1000= 7.9
12) 20÷100= .2
13) 4300÷1000= 4.3
14) .32÷10= .032
15) 980÷1000= .98
16) 150÷1000= .15
17) .03÷10= .003
18) 44÷1000= .044
19) 930÷100= 9.3
20) .79÷10= .079

PAGE: 209

Division Divide by 10, 100 and 1000 *(with decimals)*

Divide.

1) .7÷100= .007
2) 930÷100= 9.3
3) 940÷100= 9.4
4) 2.7÷10= .27
5) 43÷10= 4.3
6) 68÷1000= .068
7) 2.7÷100= .027
8) 53÷100= .53
9) .65÷10= .065
10) 63÷10= 6.3
11) 18÷10= 1.8
12) 64÷100= .64
13) 590÷100= 5.9
14) 80÷1000= .08
15) 17÷100= .17
16) 5.9÷10= .59
17) .42÷10= .042
18) .6÷10= .06
19) 90÷100= .9
20) 1.3÷10= .13

PAGE: 210

Division Divide by 10, 100 and 1000 *(with decimals)*

Divide.

1) 640÷1000= .64
2) 4.5÷100= .045
3) .67÷10= .067
4) .03÷10= .003
5) 52÷1000= .052
6) 83÷100= .83
7) .47÷10= .047
8) 960÷1000= .96
9) 9300÷1000= 9.3
10) 2900÷1000= 2.9
11) 600÷1000= .6
12) 950÷1000= .95
13) 8.8÷100= .088
14) 78÷100= .78
15) 8.5÷10= .85
16) 7.5÷100= .075
17) .85÷10= .085
18) 5200÷1000= 5.2
19) 24÷1000= .024
20) .8÷100= .008

PAGE: 211

Division Divide by 10, 100 and 1000 *(with decimals)*

Divide.

1) 22÷1000=	.022	11) 370÷1000=	.37
2) 6.5÷10=	.65	12) 670÷1000=	.67
3) 95÷100=	.95	13) 37÷10=	3.7
4) 2100÷1000=	2.1	14) 4.4÷10=	.44
5) .89÷10=	.089	15) .76÷10=	.076
6) 340÷100=	3.4	16) 32÷1000=	.032
7) 70÷100=	.7	17) 3.8÷100=	.038
8) 470÷100=	4.7	18) 83÷1000=	.083
9) 52÷1000=	.052	19) 4.6÷100=	.046
10) 500÷1000=	.5	20) 6÷1000=	.006

PAGE: 212

Division Divide by 10, 100 and 1000 *(with decimals)*

Divide.

1) 64÷10=	6.4	11) .47÷10=	.047
2) 25÷1000=	.025	12) 74÷1000=	.074
3) 90÷1000=	.09	13) 2.2÷10=	.22
4) .07÷10=	.007	14) 30÷100=	.3
5) 1.2÷10=	.12	15) 76÷100=	.76
6) 4200÷1000=	4.2	16) 570÷100=	5.7
7) 30÷100=	.3	17) 320÷1000=	.32
8) 71÷10=	7.1	18) .49÷10=	.049
9) 710÷100=	7.1	19) 5900÷1000=	5.9
10) 4400÷1000=	4.4	20) 7500÷1000=	7.5

PAGE: 213

Division Divide by 10, 100 and 1000 *(with decimals)*

Divide.

1) 290÷100=	2.9	11) 86÷100=	.86
2) .22÷10=	.022	12) 4.7÷10=	.47
3) 6.3÷10=	.63	13) 630÷1000=	.63
4) 56÷1000=	.056	14) 8700÷1000=	8.7
5) 8.7÷10=	.87	15) .48÷10=	.048
6) 540÷1000=	.54	16) 65÷100=	.65
7) 82÷10=	8.2	17) 830÷100=	8.3
8) 93÷10=	9.3	18) 750÷1000=	.75
9) .59÷10=	.059	19) 9400÷1000=	9.4
10) 4÷100=	.04	20) .43÷10=	.043

PAGE: 214

Division Divide by 10, 100 and 1000 *(with decimals)*

Divide.

1) .17÷10=	.017	11) .94÷10=	.094
2) 18÷100=	.18	12) 16÷100=	.16
3) 44÷100=	.44	13) 70÷100=	.7
4) 92÷10=	9.2	14) 67÷10=	6.7
5) 840÷1000=	.84	15) 7.1÷10=	.71
6) 83÷10=	8.3	16) 32÷100=	.32
7) 2.9÷100=	.029	17) 500÷1000=	.5
8) 5400÷1000=	5.4	18) 82÷10=	8.2
9) 170÷1000=	.17	19) 2÷100=	.02
10) 1.9÷10=	.19	20) .5÷10=	.05

PAGE: 215

Division Divide by 10, 100 and 1000 *(with decimals)*

Divide.

1) 37÷10=	3.7	11) 220÷100=	2.2
2) 44÷1000=	.044	12) 13÷10=	1.3
3) 34÷10=	3.4	13) 6200÷1000=	6.2
4) 46÷100=	.46	14) 9.5÷100=	.095
5) 93÷100=	.93	15) 470÷100=	4.7
6) 8.9÷10=	.89	16) 81÷100=	.81
7) 62÷100=	.62	17) 2800÷1000=	2.8
8) 9÷100=	.09	18) 940÷1000=	.94
9) 27÷1000=	.027	19) .08÷10=	.008
10) 1.7÷10=	.17	20) 1700÷1000=	1.7

PAGE: 216

Division Divide by 10, 100 and 1000 *(with decimals)*

Divide.

1) 2900÷1000=	2.9	11) 38÷100=	.38
2) .79÷10=	.079	12) 710÷1000=	.71
3) 58÷10=	5.8	13) 250÷1000=	.25
4) .48÷10=	.048	14) 400÷1000=	.4
5) .56÷10=	.056	15) 2÷10=	.2
6) 37÷10=	3.7	16) 50÷1000=	.05
7) 1400÷1000=	1.4	17) .27÷10=	.027
8) 110÷100=	1.1	18) 78÷10=	7.8
9) 70÷100=	.7	19) 96÷1000=	.096
10) 8.4÷10=	.84	20) .55÷10=	.055

PAGE: 218

Division Find the Missing Divisors (10, 100 or 1000)

Fill in the blanks with 10, 100 or 1000.

1) 570÷ 1000 =.57
2) 8.5÷ 100 =.085
3) 6800÷ 1000 =6.8
4) 82÷ 1000 =.082
5) 6800÷ 1000 =6.8
6) 5.1÷ 100 =.051
7) 620÷ 1000 =.62
8) 19÷ 10 =1.9
9) 3÷ 1000 =.003
10) 5400÷ 1000 =5.4
11) 42÷ 10 =4.2
12) 7600÷ 1000 =7.6
13) 63÷ 10 =6.3
14) 9400÷ 1000 =9.4
15) 1.4÷ 10 =.14
16) 290÷ 100 =2.9
17) 53÷ 10 =5.3
18) 40÷ 1000 =.04
19) 6.8÷ 10 =.68
20) 90÷ 100 =.9

PAGE: 219

Division Find the Missing Divisors (10, 100 or 1000)

Fill in the blanks with 10, 100 or 1000.

1) 24÷ 10 =2.4
2) 77÷ 100 =.77
3) .34÷ 10 =.034
4) 780÷ 100 =7.8
5) 97÷ 1000 =.097
6) 980÷ 100 =9.8
7) 340÷ 1000 =.34
8) 2.9÷ 100 =.029
9) 6÷ 100 =.06
10) 90÷ 100 =.9
11) 2.4÷ 100 =.024
12) 8.1÷ 10 =.81
13) 63÷ 100 =.63
14) 77÷ 1000 =.077
15) 66÷ 10 =6.6
16) .33÷ 10 =.033
17) 6÷ 10 =.6
18) 48÷ 100 =.48
19) 250÷ 1000 =.25
20) 78÷ 100 =.78

PAGE: 220

Division Find the Missing Divisors (10, 100 or 1000)

Fill in the blanks with 10, 100 or 1000.

1) 2.8÷ 100 =.028
2) 27÷ 1000 =.027
3) 66÷ 1000 =.066
4) 280÷ 100 =2.8
5) 380÷ 1000 =.38
6) 980÷ 100 =9.8
7) .09÷ 10 =.009
8) 830÷ 100 =8.3
9) 17÷ 10 =1.7
10) .12÷ 10 =.012
11) 84÷ 100 =.84
12) 28÷ 10 =2.8
13) 2200÷ 1000 =2.2
14) 4.6÷ 10 =.46
15) 61÷ 10 =6.1
16) 400÷ 1000 =.4
17) 83÷ 100 =.83
18) 9.1÷ 10 =.91
19) .71÷ 10 =.071
20) 55÷ 10 =5.5

PAGE: 221

Division Find the Missing Divisors (10, 100 or 1000)

Fill in the blanks with 10, 100 or 1000.

1) 370÷ 100 =3.7
2) 4.5÷ 10 =.45
3) 720÷ 100 =7.2
4) 65÷ 1000 =.065
5) 5÷ 10 =.5
6) 2.6÷ 100 =.026
7) 5600÷ 1000 =5.6
8) 29÷ 100 =.29
9) 1100÷ 1000 =1.1
10) 28÷ 100 =.28
11) 30÷ 1000 =.03
12) 150÷ 100 =1.5
13) 6.7÷ 10 =.67
14) 34÷ 10 =3.4
15) 80÷ 1000 =.08
16) .3÷ 10 =.03
17) 1400÷ 1000 =1.4
18) 6600÷ 1000 =6.6
19) 25÷ 100 =.25
20) 62÷ 100 =.62

PAGE: 222

Division Find the Missing Divisors (10, 100 or 1000)

Fill in the blanks with 10, 100 or 1000.

1) 93÷ 100 =.93
2) 2÷ 100 =.02
3) .84÷ 10 =.084
4) 5.5÷ 100 =.055
5) 90÷ 100 =.9
6) 7.3÷ 100 =.073
7) 4.9÷ 10 =.49
8) 180÷ 100 =1.8
9) 29÷ 10 =2.9
10) 3400÷ 1000 =3.4
11) 11÷ 1000 =.011
12) 64÷ 100 =.64
13) 67÷ 1000 =.067
14) 360÷ 100 =3.6
15) 460÷ 100 =4.6
16) 52÷ 1000 =.052
17) .41÷ 10 =.041
18) 58÷ 1000 =.058
19) 7.1÷ 100 =.071
20) 1.8÷ 100 =.018

PAGE: 223

Division Find the Missing Divisors (10, 100 or 1000)

Fill in the blanks with 10, 100 or 1000.

1) 88÷ 100 =.88
2) 8700÷ 1000 =8.7
3) 810÷ 100 =8.1
4) 80÷ 1000 =.08
5) 530÷ 1000 =.53
6) 6.8÷ 10 =.68
7) 940÷ 1000 =.94
8) 4.6÷ 100 =.046
9) .31÷ 10 =.031
10) 79÷ 10 =7.9
11) 3200÷ 1000 =3.2
12) .7÷ 10 =.07
13) 3.7÷ 100 =.037
14) .13÷ 10 =.013
15) .73÷ 10 =.073
16) 3.2÷ 100 =.032
17) 120÷ 100 =1.2
18) 45÷ 1000 =.045
19) 98÷ 1000 =.098
20) .04÷ 10 =.004

PAGE: 224

Division Find the Missing Divisors (10, 100 or 1000)

Fill in the blanks with 10, 100 or 1000.

1) 440÷ 100 =4.4	11) 6.5÷ 100 =.065
2) 320÷ 100 =3.2	12) 900÷ 1000 =.9
3) 600÷ 1000 =.6	13) 93÷ 10 =9.3
4) 930÷ 100 =9.3	14) 5800÷ 1000 =5.8
5) 920÷ 100 =9.2	15) 2÷ 100 =.02
6) 390÷ 1000 =.39	16) 9.5÷ 10 =.95
7) 59÷ 100 =.59	17) 10÷ 1000 =.01
8) 5.4÷ 100 =.054	18) 2300÷ 1000 =2.3
9) 9.8÷ 10 =.98	19) 820÷ 100 =8.2
10) 48÷ 1000 =.048	20) .02÷ 10 =.002

PAGE: 225

Division Find the Missing Divisors (10, 100 or 1000)

Fill in the blanks with 10, 100 or 1000.

1) 2÷ 100 =.02	11) 1.5÷ 100 =.015
2) 18÷ 100 =.18	12) .57÷ 10 =.057
3) 8800÷ 1000 =8.8	13) .87÷ 10 =.087
4) 6900÷ 1000 =6.9	14) 11÷ 1000 =.011
5) 93÷ 1000 =.093	15) 96÷ 1000 =.096
6) 110÷ 1000 =.11	16) 240÷ 1000 =.24
7) 930÷ 100 =9.3	17) 870÷ 1000 =.87
8) 13÷ 1000 =.013	18) 170÷ 1000 =.17
9) 31÷ 1000 =.031	19) 1.1÷ 100 =.011
10) 44÷ 1000 =.044	20) 3900÷ 1000 =3.9

PAGE: 226

Division Find the Missing Divisors (10, 100 or 1000)

Fill in the blanks with 10, 100 or 1000.

1) 8.4÷ 10 =.84	11) 29÷ 1000 =.029
2) 3.3÷ 100 =.033	12) 8800÷ 1000 =8.8
3) 37÷ 1000 =.037	13) 450÷ 1000 =.45
4) 45÷ 100 =.45	14) 480÷ 1000 =.48
5) .27÷ 10 =.027	15) 1500÷ 1000 =1.5
6) 95÷ 10 =9.5	16) 4.1÷ 100 =.041
7) 43÷ 10 =4.3	17) .38÷ 10 =.038
8) 5.2÷ 100 =.052	18) 4.7÷ 100 =.047
9) 290÷ 100 =2.9	19) 7÷ 100 =.07
10) 28÷ 100 =.28	20) 9.3÷ 100 =.093

PAGE: 227

Division Find the Missing Divisors (10, 100 or 1000)

Fill in the blanks with 10, 100 or 1000.

1) 5.7÷ 100 =.057	11) 1800÷ 1000 =1.8
2) 790÷ 100 =7.9	12) 1.9÷ 100 =.019
3) 60÷ 1000 =.06	13) 9400÷ 1000 =9.4
4) 34÷ 10 =3.4	14) 42÷ 10 =4.2
5) 560÷ 100 =5.6	15) .9÷ 10 =.09
6) 330÷ 1000 =.33	16) .31÷ 10 =.031
7) 3.8÷ 100 =.038	17) 5.6÷ 10 =.56
8) .43÷ 10 =.043	18) 6.4÷ 100 =.064
9) 4300÷ 1000 =4.3	19) 930÷ 1000 =.93
10) 78÷ 100 =.78	20) 70÷ 100 =.7

PAGE: 228

Division Find the Missing Divisors (10, 100 or 1000)

Fill in the blanks with 10, 100 or 1000.

1) 3.7÷ 100 =.037	11) 1.7÷ 100 =.017
2) .8÷ 10 =.08	12) 36÷ 100 =.36
3) 8400÷ 1000 =8.4	13) 56÷ 1000 =.056
4) 2500÷ 1000 =2.5	14) .59÷ 10 =.059
5) 43÷ 10 =4.3	15) 74÷ 10 =7.4
6) 9.4÷ 100 =.094	16) .7÷ 10 =.07
7) 1800÷ 1000 =1.8	17) 280÷ 1000 =.28
8) 5300÷ 1000 =5.3	18) .13÷ 10 =.013
9) 84÷ 1000 =.084	19) 55÷ 1000 =.055
10) 5.1÷ 100 =.051	20) 75÷ 1000 =.075

Made in the USA
Monee, IL
21 September 2020

43142348R00149